Electromagnetics and Antenna Optimization Using Taguchi's Method

Electromagnetics and Antenna Optimization Using Taguchi's Method

Wei-Chung Weng, Fan Yang and Atef Elsherbeni

ISBN: 978-3-031-00573-2   paperback

ISBN: 978-3-031-01701-8   ebook

DOI: 10.1007/978-3-031-01701-8

A Publication in the Springer series

SYNTHESIS LECTURES ON COMPUTATIONAL ELECTROMAGNETICS #18

Lecture #18

Series Editor: Constantine A. Balanis, Arizona State University

Series ISSN

ISSN 1932-1252 print

ISSN 1932-1716 electronic

Electromagnetics and Antenna Optimization Using Taguchi's Method

Wei-Chung Weng
National Chi Nan University, Taiwan

Fan Yang and Atef Elsherbeni
The University of Mississippi, Oxford, Mississippi

SYNTHESIS LECTURES ON COMPUTATIONAL ELECTROMAGNETICS #18

ABSTRACT

This book presents a new global optimization technique using Taguchi's method and its applications in electromagnetics and antenna engineering. Compared with traditional optimization techniques, Taguchi's optimization method is easy to implement and very efficient in reaching optimum solutions.

Taguchi's optimization method is developed based on the orthogonal array (OA) concept, which offers a systematic and efficient way to select design parameters. The book illustrates the basic implementation procedure of Taguchi's optimization method and discusses various advanced techniques for performance improvement. In addition, the integration of Taguchi's optimization method with commercial electromagnetics software is introduced in the book.

The proposed optimization method is used in various linear antenna arrays, microstrip filters, and ultra-wideband antenna designs. Successful examples include linear antenna array with a null controlled pattern, linear antenna array with a sector beam, linear antenna array with reduced side lobe levels, microstrip band stop filter, microstrip band pass filter, coplanar waveguide band stop filter, coplanar ultra-wide band antenna, and ultra-wide band antenna with band notch feature.

Satisfactory results obtained from the design process demonstrate the validity and efficiency of the proposed Taguchi's optimization method.

KEYWORDS

antenna array, microwave filter, optimization method, orthogonal array, Taguchi's method, ultra-wideband antenna.

Contents

CHAPTER 1

Introduction

1.1 BACKGROUND

The purpose of optimization is to try to achieve the best result [1]. By adjusting input parameters, the process of optimization seeks for a better output so that the performance of a system, such as quality, monetary cost, and efficiency, can be improved.

Optimization can be applied to a variety of areas and has received great attention recently. Thanks to the rapid development of computer technology, many optimization techniques such as genetic algorithm (GA), particle swarm optimization (PSO), simulated annealing (SA), artificial neural network (ANN), and gradient-based techniques have been implemented by computer codes. Currently in the field of electromagnetics (EM), many microwave circuits and antenna designs rely on optimization techniques [2–4]. Traditional methods, such as the trial-and-error approach, require many experiments to obtain an optimum or a satisfactory result. Therefore, an optimization technique is necessary for EM applications.

In general, optimization methods can be divided into two categories: global and local techniques. Global techniques have several advantages. For example, their solutions are largely independent of initial conditions. In addition, they are capable of handling discontinuous and nondifferentiable objective functions. Furthermore, global techniques exhibit good performance when dealing with solution spaces that have discontinuities, constrained parameters, and a large number of dimensions with many potential local maxima. However, a main drawback is that the convergence rate is slow. In contrast, for the local techniques, the main advantage is that the solution converges rapidly. However, local techniques depend highly on the starting point or initial guess. In addition, local techniques react poorly to the presence of discontinuities in solution spaces.

Table 1.1 shows the characteristics comparison of various optimization techniques [5] that are commonly used in the EM field. GA, SA, PSO, and Taguchi's method are considered as global optimization methods while gradient-based methods are local optimization methods.

In EM applications, global methods are favored over local methods. Global techniques yield either a global or near global optimum instead of a local optimum. They often find useful solutions when other local techniques fail. Global methods are particularly useful when dealing with new problems in which the nature of the solution space is relatively unknown.

TABLE 1.1: Characteristics comparison of optimization techniques

METHOD	CHARACTERISTICS			
	GLOBAL OPTIMIZATION	DISCONTINUOUS FUNCTION	NONDIFFERENTIABLE FUNCTION	CONVERGENCE RATE
Gradient-based	Poor	Poor	Poor	Good
Random	Fair	Good	Good	Poor
GA	Good	Good	Good	Fair
PSO	Good	Good	Good	Good
Taguchi's method	Good	Good	Good	Good
ANN	Fair	Good	Good	Good
SA	Good	Good	Good	Fair

1.2   TAGUCHI'S METHOD

1.2.1   Objective

In general, experiments are used to study the performance of systems or processes. According to the results of the current experiment, one may adjust the values of system parameters in the next experiment to achieve a better performance, which is called *a trial-and-error approach*. The drawback of this strategy arises when the obtained result is not the optimum or the system requirements cannot be satisfied after a large number of experiments. Alternatively, researchers may want to test all combinations of parameters in an experiment, which is called *a full factorial experiment*. This strategy can cover all possibilities in the experiment and determine the optimal result. However, it will run too many trials and hence cost much time and money in practice.

To solve the above difficulties, Taguchi's method was developed based on the concept of the orthogonal array (OA), which can effectively reduce the number of tests required in a design process [6]. It provides an efficient way to choose the design parameters in an optimization procedure. Although Taguchi's method has been successfully applied in many fields such as chemical engineering, mechanical engineering, integrated chip manufacture, and power electronics [7–10], it is not well known in the EM community. Only limited applications are available for design of absorbers [11], [12], electrically conductive adhesives [13], diplexers [14], statistical characterization of microwave circuit parameters [15], and linear antenna array synthesis [16], [17]. A major goal of this book is to further expose Taguchi's method to the EM community and to demonstrate its great potential in EM optimizations.

Based on the authors' experience, Taguchi's method has the following advantages:

- simple implementation;
- effective reduction of test trials;
- fast convergence speed;
- global optimum results;
- independence from initial values of optimization parameters.

1.2.2   Research Work

This book covers both the algorithm development and EM applications of Taguchi's method. Figure 1.1 depicts the conceptual flow graph used for the organization of this book. The following notes give some introductions to each of the remaining chapters.

Orthogonal arrays play an essential role in Taguchi's method. The fundamental concepts of OAs are presented in Chapter 2. For certain applications, one may want to create an OA by a computer code. Therefore, an OA construction algorithm that can construct an arbitrary odd-level and two-strength OA is presented in Chapter 2 as well.

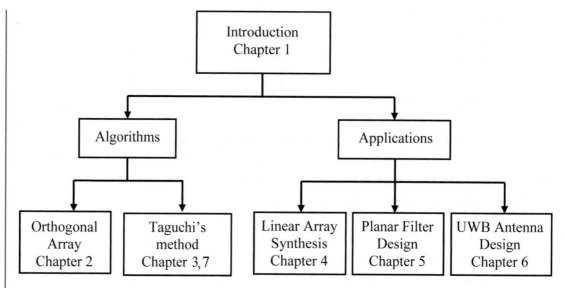

FIGURE 1.1: Conceptual flow graph for the organization of this book.

The detailed implementation procedure of Taguchi's method is presented in Chapter 3. To show the global optimization ability of Taguchi's method, complex two-dimensional test functions with many local maxima/minima are used as examples. Optimization results demonstrate the excellent performance of searching for the global optimum. Some further improvement techniques of Taguchi's methods are also presented in Chapter 3. In addition, the OA concept can be used to enhance the performance of other optimization approaches such as PSO [18]. It is demonstrated in Chapter 7 that the optimization efficiency of OA-PSO is better than the classical PSO.

Following the algorithm are the EM applications of Taguchi's optimization method. Linear antenna array optimization has received great attention in the EM community. Recently, GA and PSO have been successfully applied in designing linear antenna arrays [19–25]. Besides GA and PSO methods, this research uses Taguchi's method to design linear antenna arrays that produce a null controlled radiation pattern, a sector beam radiation pattern [16], and a suppressed side lobe level (SLL) pattern. A detailed implementation procedure is presented, and each step is illustrated by the array example shown in Chapter 4. This study shows that the proposed method is straightforward and easy to implement, and can quickly converge to the optimum designs.

Microwave filters are widely used in telecommunication devices. For example, a band-stop filter (BSF) is used to suppress the noises and interference signals coming from the environments, while a band-pass filter (BPF) allows desired signal frequencies to pass through the communication channels. At low frequencies, filters are usually designed with lumped elements to realize the desired frequency responses. At microwave-frequency range, operating wavelength is comparable to the circuit dimensions. Therefore, passive printed types of filters are usually used in microwave applications. These filters are planar structured and composed of several stubs of microstrip lines. They need to be accurately modeled using full-wave EM simulators. In this research, a full-wave simulator, IE3D [26], is used to analyze the performance of filters while Taguchi's method is applied as an external optimizer to drive IE3D and optimize those microstrip filters. The efficiency of this design procedure verifies the excellent optimization performance of Taguchi's method, and the optimized results are shown in Chapter 5.

Taguchi's method is also used in ultra-wideband antenna designs. In 2002, the Federal Communications Commission (FCC) released the ultra-wideband (UWB) system specification [27] whose spectrum covers from 3.1 to 10.6 GHz. Planar UWB antennas are desirable because of advantages such as low profile, light weight, low cost, and easy fabrication. The advantages of a coplanar waveguide (CPW) over a microstrip line are low radiation loss, balanced line, low dispersion, same conducting plane, and no need for vias. Therefore, a planar UWB antenna fed by a CPW line is designed in this study. A full-wave commercial simulator, IE3D, along with an external Taguchi method-based optimizer is used to optimize a UWB antenna. In Chapter 6, the results of the optimized antenna show that the impedance bandwidth is from 3 to 12.1 GHz, which not only can cover the UWB spectrum but also can be used for X band radar applications. Furthermore, since the spectrum of wireless local network (WLAN) of the *IEEE* 802.11a standard is located between 5.15 and 5.825 GHz, it is desirable that a UWB antenna has a band notch feature at the center frequency (5.5 GHz) to avoid potential interference between the UWB and WLAN bands. In this study, a compact BSF is integrated in the UWB antenna design to achieve a stop band between 5 and 6 GHz without changing the antenna's geometry and frequency responses outside the stop band.

At the end of the book, a brief summary is presented and several future research directions are suggested for the readers. It is the authors' hope that this book will help readers with similar research interests and stimulate the future development in EM optimizations and antenna designs.

• • • •

CHAPTER 2

Orthogonal Arrays

The development of Taguchi's method is based on orthogonal arrays (OAs) that have a profound background in statistics [28]. Orthogonal arrays were introduced in the 1940s and have been widely used in designing experiments. They provide an efficient and systematic way to determine control parameters so that the optimal result can be found with only a few experimental runs. This chapter briefly reviews the fundamental concepts of OAs, such as their definition, important properties, and constructions.

2.1 DEFINITION OF ORTHOGONAL ARRAY

Definition: Let S be a set of s symbols or levels. A matrix A of N rows and k columns with entries from S is said to be an *OA with s levels and strength t* ($0 \leq t \leq k$) if in every $N \times t$ subarray of A, each t-tuple based on S appears exactly the same times as a row [28, Chapter 1]. The notation $OA(N, k, s, t)$ is used to represent an OA.

To help readers understand the OA definition, Table 2.1 shows an orthogonal array OA(*27, 10, 3, 2*), which has 27 rows and 10 columns. Each entry of the array is selected from a set $S = \{1, 2, 3\}$. Thus, this is a three-level OA. Pick any arbitrary two columns ($t = 2$) and one may see nine possible combinations as a row: (1, 1), (1, 2), (1, 3), (2, 1), (2, 2), (2, 3), (3, 1), (3, 2), (3, 3). It can be simply verified that each combination has the same number of occurrences as a row, i.e., three times. This is the meaning of "orthogonal" in the definition, which ensures a balanced and fair selection of parameters in all possible combinations.

When this OA is used to design experiments, the 10 columns represent 10 parameters that need to be optimized. For each column, the entries 1, 2, and 3 denote three specific statuses or levels that an optimization parameter may select from. Note that for different optimization parameters, the levels 1, 2, and 3 may correspond to different numerical values. For example, if the optimization range for the parameter 1 is [0, 1], the corresponding values for levels (1, 2, 3) could be (0.25, 0.5, 0.75). In contrast, if the optimization range for the parameter 2 is [−1, 0], the corresponding values for levels (1, 2, 3) change to (−0.75, −0.5, −0.25). Therefore, the corresponding values for the levels depend on the parameters and vary in different applications.

Each row of the OA describes a certain combination of the levels for these 10 parameters. For example, the first row means that all parameters take the level 1. The second row means that

EXPERIMENTS	ELEMENTS									
	1	2	3	4	5	6	7	8	9	10
1	1	1	1	1	1	1	1	1	1	1
2	2	1	2	2	2	3	3	1	2	3
3	3	1	3	3	3	2	2	1	3	2
4	1	2	1	2	2	2	3	3	1	2
5	2	2	2	3	3	1	2	3	2	1
6	3	2	3	1	1	3	1	3	3	3
7	1	3	1	3	3	3	2	2	1	3
8	2	3	2	1	1	2	1	2	2	2
9	3	3	3	2	2	1	3	2	3	1
10	1	1	2	1	2	2	2	3	3	1
11	2	1	3	2	3	1	1	3	1	3
12	3	1	1	3	1	3	3	3	2	2
13	1	2	2	2	3	3	1	2	3	2
14	2	2	3	3	1	2	3	2	1	1
15	3	2	1	1	2	1	2	2	2	3
16	1	3	2	3	1	1	3	1	3	3
17	2	3	3	1	2	3	2	1	1	2
18	3	3	1	2	3	2	1	1	2	1
19	1	1	3	1	3	3	3	2	2	1
20	2	1	1	2	1	2	2	2	3	3
21	3	1	2	3	2	1	1	2	1	2
22	1	2	3	2	1	1	2	1	2	2
23	2	2	1	3	2	3	1	1	3	1
24	3	2	2	1	3	2	3	1	1	3
25	1	3	3	3	2	2	1	3	2	3
26	2	3	1	1	3	1	3	3	3	2
27	3	3	2	2	1	3	2	3	1	1

TABLE 2.1: The OA(*27, 10, 3, 2*)

parameters 2 and 8 take level 1; parameters 1, 3, 4, 5, and 9 take level 2; and parameters 6, 7, and 10 take level 3. Once each parameter is assigned to a corresponding level value, one can conduct the experiment and find the output result. It is important to point out that the 27 rows of the OA indicate that 27 experiments need to be carried out per design iteration.

2.2   IMPORTANT PROPERTIES

The OAs have several important properties. For the brevity of the book, three fundamental characteristics are highlighted here, which are useful for Taguchi's method discussed in later chapters. The first one is the fractional factorial characteristic. Using the above example that includes 10 parameters where each has three levels, one notices that a full factorial strategy needs to conduct $3^{10} = 59\,049$ experiments. In contrast, if one uses the OA to design such experiments, only 27 experiments are needed. After a simple analysis and processing of the output results from the 27 experiments, an optimum combination of the parameter values can be obtained [6, Chapter 15]. Although the number of experiments is dramatically reduced from 59 049 to 27, statistical results demonstrate that the optimum outcome obtained from the OA usage is close to that obtained from the full-factorial approach.

The second fundamental property of the OA is that all possible combinations of up to t parameters occur equally, which ensures a balanced and fair comparison of levels for *any parameter* and *any interactions of parameters*. A quick examination of Table 2.1 reveals that for each parameter (column), levels 1, 2, and 3 have nine times of occurrences. Thus, all possible levels of a parameter are tested equally. A similar property applies for the combination of any two parameters. Therefore, the OA approach investigates not only the effects of the individual parameters on the experiment outcome, but also the interactions of any two parameters.

In general, one could increase the strength t of the OA to consider interactions between more parameters. However, the larger the strength t is, the more rows (experiments) the OA has. The OAs used in this book have a strength of 2, which is found to be efficient for the problems considered.

The third useful property of OAs is that any $N \times k'$ subarray of an existing OA(N, k, s, t) is still an OA with a notation OA(N, k', s, t'), where $t' = \min\{k', t\}$. In other words, if one or more columns are deleted from an OA, the resulting array is still an OA but with a smaller number of parameters. For example, if we delete the last two columns in Table 2.1, we can obtain an OA($27, 8, 3, 2$). This property is especially useful when selecting an OA from an existing OA database. If an OA with a certain number of columns (k') cannot be found in the database, one can choose an OA with a larger number of columns ($k > k'$) and manually delete the redundant ($k - k'$) columns to obtain the desired OA.

More OA properties can be found in the literature and interested readers are suggested to read [28] for more information.

2.3   EXISTENCE AND CONSTRUCTION OF ORTHOGONAL ARRAYS

Two fundamental questions regarding OAs, namely, existence and construction, are discussed in this section. A simple statement of the existence question is: for given values of k, s, and t, determine the minimum number of rows N so that an OA(N, k, s, t) exists. The reason one would like to find the minimum N is that a small number of experiments is preferred in practice. Rao's [29] inequalities are established to answer the existence question. The parameters of OA(N, k, s, t) must satisfy the following inequalities:

$$N \geq \sum_{i=0}^{u} \binom{k}{i} (s-1)^i, \text{ if } t = 2u, \quad u > 0, \tag{2.1}$$

$$N \geq \sum_{i=0}^{u} \binom{k}{i} (s-1)^i + \binom{k-1}{u} (s-1)^{u+1}, \text{ if } t = 2u+1, \ u \geq 0. \tag{2.2}$$

Further improvements on Rao's bound can be found in [28] for OAs with strength 2 and 3 ($t = 2, 3$) and for two-level OAs ($s = 2$).

The second question is how to construct an OA(N, k, s, t). Numerous techniques are known for constructing OAs. Galois fields [28, Chapter 3] turn out to be a powerful tool for the construction of OAs, and several methods are proposed using such fields and finite geometries. In addition, there is a close relation between OAs and coding theory. Thus, many construction techniques for OAs are proposed based on error-correcting codes. Furthermore, the difference scheme is also known as one of the earliest methods to construct certain OAs. Nowadays, many OAs with different numbers of parameters, levels, and strengths have been developed and archived in OA databases or libraries, which can be found in books related to OA or Taguchi's method. The OAs used in this book are listed in [30].

For certain applications, one may want to create an OA automatically by a computer code. An OA construction algorithm was presented in [31], which can construct an arbitrary *odd-level* ($s = 3$, $5, 7, \ldots$) and *two-strength* ($t = 2$) OA. For an OA(N, k, s, t), the values of N and k can be determined by the following two equations:

$$N = s^p, \tag{2.3}$$

$$k = \frac{N-1}{s-1}, \tag{2.4}$$

where p is a positive integer number starting with 2, namely, $p = 2, 3, \ldots$. The OA construction algorithm can be divided into two parts: *construction of basic columns* and *construction of non-basic columns*. Non-basic columns are linear combinations of basic columns. The pseudo-code of the algorithm is as follows:

1) *Construct basic columns*:

   for $ii = 1$ to p

   $$k = \frac{s^{ii} - 1}{s - 1}$$

   for $m = 1$ to N

   $$a(m, k) = \left\lfloor \frac{m - 1}{s^p - 1} \right\rfloor \bmod s$$

   end   (loop m)

   end   (loop ii)

2) *Construct non-basic columns*:

   for $ii = 2$ to p

   $$k = \frac{s^{ii} - 1}{s - 1} + 1$$

   for $jj = 1$ to k-1

   for $kk = 1$ to s-1

   $$a(1 \sim m, k + kk + (jj - 1)(s - 1)) = (a(1 \sim m, jj) \; kk + a(1 \sim m, k)) \bmod s$$

   end   (loop kk)

   end   (loop jj)

   end   (loop ii)

   $$a(m, k) = a(m, k) + 1$$

The OA(*9, 4, 3, 2*) and OA(*25, 6, 5, 2*) are constructed by this OA construction code and shown in Tables 2.2 and 2.3, respectively. The grey background indicates the basic columns of the OA while the rest are non-basic columns.

A sample Matlab code is attached below for users' reference.

% This OA construction Matlab can construct an arbitrary odd-level, two-strength OA.

% User should change the value of S and J which are indicated by "*". After execution of

% this code, the OA can be obtained in the current directory by the file name

```
% "OA(N, K, S, 2).txt" . This code was designed by Wei-Chung Weng, February 2007.
% All rights reserved.

clc; clear all;
S = 3;  %level " * "
J = 2;  % related to M  " * "
M = S^J; % # of experiment

for k = 1:J  %  for basic columns
    j = (S^(k-1)-1)/(S-1) + 1;

    for i = 1:M
        A(i,j) = mod(floor((i-1)/(S^(J-k))),S);
    end
end
for k = 2:J  %  for non-basic columns
    j = (S^(k-1)-1)/(S-1) + 1;

    for p = 1:j-1
        for q = 1:S-1
            A(:,(j + (p-1)*(S-1) + q)) = mod((A(:,p)*q + A(:,j)),S);
        end
    end
end

% A = A + 1;   % if values start from 1

  % output data

[N,K] = size(A);
str1 = num2str(N,'%0.1d');
str2 = num2str(K,'%0.1d');
str3 = num2str(S,'%0.1d');
TT = ['OA(' str1 ',' str2 ',' str3 ', 2).txt'];
fid2 = fopen(TT,'wt');
```

```
for j = 1:N
  for k = 1:K
    fprintf(fid2,'%0.1d   ',A(j,k));
    if k = = K
      fprintf(fid2,'\n');
    end
  end
end
fclose(fid2);
```

TABLE 2.2: The OA(*9, 4, 3, 2*) is constructed by the OA construction code (*s* = 3, *p* = 2, *N* = 9, *k* = 4)

EXPERIMENT	ELEMENT			
	1	2	3	4
1	1	1	1	1
2	1	2	2	2
3	1	3	3	3
4	2	1	2	3
5	2	2	3	1
6	2	3	1	2
7	3	1	3	2
8	3	2	1	3
9	3	3	2	1

TABLE 2.3: The OA(*25, 6, 5, 2*) is constructed by the OA construction code ($s = 5$, $p = 2$, $N = 25$, $k = 6$)

EXPERIMENT	ELEMENT					
	1	2	3	4	5	6
1	1	1	1	1	1	1
2	1	2	2	2	2	2
3	1	3	3	3	3	3
4	1	4	4	4	4	4
5	1	5	5	5	5	5
6	2	1	2	3	4	5
7	2	2	3	4	5	1
8	2	3	4	5	1	2
9	2	4	5	1	2	3
10	2	5	1	2	3	4
11	3	1	3	5	2	4
12	3	2	4	1	3	5
13	3	3	5	2	4	1
14	3	4	1	3	5	2
15	3	5	2	4	1	3
16	4	1	4	2	5	3
17	4	2	5	3	1	4
18	4	3	1	4	2	5
19	4	4	2	5	3	1
20	4	5	3	1	4	2
21	5	1	5	4	3	2
22	5	2	1	5	4	3
23	5	3	2	1	5	4
24	5	4	3	2	1	5
25	5	5	4	3	2	1

CHAPTER 3

Taguchi's Optimization Method

This chapter presents the optimization technique using Taguchi's method. A fundamental implementation procedure is introduced first and several optimization problems are used as examples to demonstrate its validity. Based on this development, several improvement techniques are proposed to enhance the optimization performance.

3.1   IMPLEMENTATION PROCEDURE

This section introduces a novel iterative procedure of Taguchi's optimization method, as shown in Figure 3.1. To illustrate the implementation procedure, a 10-dimensional Rastigrin function [32]–[33] is used as an example and listed in (3.1).

$$f(x) = \sum_{i=1}^{10} [x_i^2 - 10\cos(2\pi x_i) + 10], \quad -9 < x_i < 8 \qquad (3.1)$$

The function has a global minimum 0 when all $x_i = 0$. The 10 input parameters $(x_i, i = 1, \ldots, 10)$ are optimized to find the global minimum. The search range is set to [−9, 8] to obtain an asymmetrical optimization space. The optimization procedure of the first iteration is explained in detail, and the procedure of the remaining iterations is similar to that of the first iteration.

3.1.1   Problem Initialization

The optimization procedure starts with the problem initialization, which includes the *selection of a proper OA* and *the design of a suitable fitness function*. The selection of an OA(N, k, s, t) mainly depends on the number of optimization parameters. In (3.1), there are 10 parameters that should be optimized. Thus, the OA to be selected must have 10 columns (k = 10) to represent these parameters. To characterize the nonlinear effect, three levels (s = 3) are found sufficient for each input parameter. Usually, an OA with a strength of 2 (t = 2) is efficient for most problems because it results in a small number of rows in the array. In summary, an OA with 10 columns, 3 levels, and 2 strengths is needed.

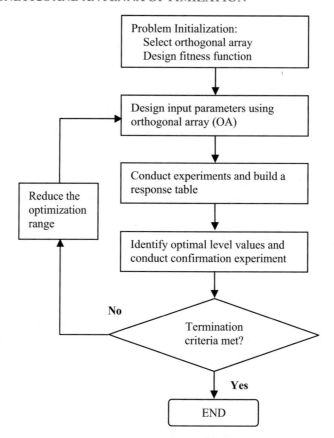

FIGURE 3.1: Flowchart of Taguchi's optimization method. From [17], copyright © IEEE 2007.

After searching the online OA database [30], an OA(*27, 13, 3, 2*) is found to be available. To be consistent with this problem, the first 10 columns of the OA are retained whereas the rest (3 columns) are deleted. Hence, an OA(*27, 10, 3, 2*) is obtained for the optimization process, which is shown in Table 2.1.

The fitness function is chosen according to the optimization goal. In this optimization example the fitness function is selected to be the same as the Rastigrin function:

$$\text{Fitness} = f(x) = \sum_{i=1}^{10} [x_i^2 - 10\cos(2\pi x_i) + 10], \qquad (3.2)$$

where the fitness can be considered as the difference between the optimization goal (0 value) and the obtained value from the current inputs x_i. The smaller the fitness value, the better the match between the obtained value and the desired one.

3.1.2 Design Input Parameters Using OA

Next, the input parameters need to be selected to conduct the experiments. When the OA is used, the corresponding numerical values for the three levels of each input parameter should be determined.

In the first iteration, the value for level 2 is selected at the center of the optimization range. Values of levels 1 and 3 are calculated by subtracting/adding the value of level 2 with a variable called *level difference* (LD). The *level difference* in the first iteration (LD_1) is determined by the following equation:

$$LD_1 = \frac{\max - \min}{\text{number of levels} + 1} = \frac{8 - (-9)}{3 + 1} = 4.25, \tag{3.3}$$

where max is the upper bound of the optimization range and min is the lower bound of the optimization range. Thus, the three levels are uniformly distributed in the optimization region. With the use of (3.3), each entry of the OA in Table 2.1 can be converted into a corresponding level value of the input parameter $x_n|_1^m$, as shown in Table 3.1, where n indicates the nth element, the subscript 1 indicates the first iteration, and the superscript m indicates the level 1, 2, or 3.

3.1.3 Conduct Experiments and Build a Response Table

After determining the input parameters, the fitness function for each experiment can be calculated. For example, the fitness value for experiment 1 with all parameters being level 1 is computed using (3.2) and the result is 325.63. Next, the fitness value is converted to the signal-to-noise (S/N) ratio (η) in Taguchi's method using the following formula

$$\eta = -20 \log (\text{Fitness}) \tag{3.4}$$

Hence, a small fitness value results in a large S/N ratio. After conducting all experiments in the first iteration, the fitness values and corresponding S/N ratios are obtained and listed in Table 3.1.

These results are then used to build a response table for the first iteration by averaging the S/N ratios for each parameter n and each level m using the following equation:

$$\overline{\eta}(m, n) = \frac{s}{N} \sum_{i, OA(i,n) = m} \eta_i. \tag{3.5}$$

For example, the average S/N ratio for $x_7|_1^2$ is

$$\overline{\eta}(2,7) = \frac{1}{9} \sum_{i,\ OA(i,7)=2} \eta_i = \frac{1}{9}[(-47.83) + (-47.55) + (-48.25) + (-48.28) + (-47.86)$$
$$+ (-47.99) + (-47.86) + (-48.15) + (-48.12)] = -47.99\ (dB).$$

Therefore, the response table is created and is shown in Table 3.2.

TABLE 3.1: The OA(27, 10, 3, 2), level values, fitness values, and S/N ratios in the first iteration of the Rastigrin function optimization

EXPERIMENTS	ELEMENTS										FITNESS	S/N RATIO (dB), η
	1	2	3	4	5	6	7	8	9	10		
1	-4.75	-4.75	-4.75	-4.75	-4.75	-4.75	-4.75	-4.75	-4.75	-4.75	325.63	-50.25
2	-0.5	-4.75	-0.5	-0.5	-0.5	3.75	3.75	-4.75	-0.5	3.75	238.56	-47.55
3	3.75	-4.75	3.75	3.75	3.75	-0.5	-0.5	-4.75	3.75	-0.5	246.19	-47.83
4	-4.75	-0.5	-4.75	-0.5	-0.5	-0.5	3.75	3.75	-4.75	-0.5	247.06	-47.86
5	-0.5	-0.5	-0.5	3.75	3.75	-4.75	-0.5	3.75	-0.5	-4.75	238.56	-47.55
6	3.75	-0.5	3.75	-4.75	-4.75	3.75	-4.75	3.75	3.75	3.75	262.31	-48.38
7	-4.75	3.75	-4.75	3.75	3.75	3.75	-0.5	-0.5	-4.75	3.75	258.5	-48.25
8	-0.5	3.75	-0.5	-4.75	-4.75	-0.5	-4.75	-0.5	-0.5	-0.5	243.25	-47.72
9	3.75	3.75	3.75	-0.5	-0.5	-4.75	3.75	-0.5	3.75	-4.75	246.19	-47.83
10	-4.75	-4.75	-4.75	-0.5	-4.75	-4.75	3.75	3.75	-4.75	-4.75	259.38	-48.28
11	-0.5	-4.75	-0.5	3.75	-4.75	3.75	-4.75	3.75	-0.5	3.75	267.00	-48.53
12	3.75	-4.75	3.75	-0.5	3.75	3.75	3.75	3.75	-0.5	3.75	258.50	-48.25
13	-4.75	-0.5	-0.5	-0.5	3.75	3.75	-4.75	-0.5	3.75	-0.5	238.56	-47.55
14	-0.5	-0.5	3.75	3.75	-4.75	-0.5	3.75	-0.5	-4.75	-4.75	250.88	-47.99

TABLE 3.1: Continued

EXPERIMENTS	ELEMENTS										FITNESS	S/N RATIO (dB), η
	1	2	3	4	5	6	7	8	9	10		
15	3.75	-0.5	-4.75	-4.75	-0.5	-4.75	-0.5	-0.5	-0.5	3.75	247.06	-47.86
16	-4.75	3.75	-0.5	3.75	-4.75	-4.75	3.75	-4.75	3.75	3.75	270.81	-48.65
17	-0.5	3.75	3.75	-4.75	-0.5	3.75	-0.5	-4.75	-4.75	-0.5	250.88	-47.99
18	3.75	3.75	-4.75	-0.5	3.75	-0.5	-4.75	-4.75	-0.5	-4.75	263.19	-48.41
19	-4.75	-4.75	3.75	-4.75	3.75	3.75	3.75	-0.5	-0.5	-4.75	267.00	-48.53
20	-0.5	-4.75	-4.75	-0.5	-4.75	-0.5	-0.5	-0.5	3.75	3.75	247.06	-47.86
21	3.75	-4.75	-0.5	3.75	-0.5	-4.75	-4.75	-0.5	-4.75	-0.5	259.38	-48.28
22	-4.75	-0.5	3.75	-0.5	-4.75	-4.75	-0.5	-4.75	-0.5	-0.5	255.56	-48.15
23	-0.5	-0.5	-4.75	3.75	-0.5	3.75	-4.75	-4.75	3.75	-4.75	263.19	-48.41
24	3.75	-0.5	-0.5	-4.75	3.75	-0.5	3.75	-4.75	-4.75	3.75	254.69	-48.12
25	-4.75	3.75	3.75	3.75	-0.5	-0.5	-4.75	3.75	-0.5	3.75	246.19	-47.83
26	-0.5	3.75	-4.75	-4.75	3.75	-4.75	3.75	3.75	3.75	-0.5	258.50	-48.25
27	3.75	3.75	-0.5	-0.5	-4.75	3.75	-0.5	3.75	-4.75	-4.75	254.69	-48.12

TABLE 3.2: Response table after the first iteration for the Rastigrin function optimization (decibel)

LEVELS	ELEMENTS									
	1	2	3	4	5	6	7	8	9	10
1	−48.37	−48.37	−48.38	−48.38	−48.37	−48.37	−48.37	−48.37	−48.38	−48.37
2	−47.98	−47.98	−47.98	−47.98	−47.99	−47.99	−47.99	−47.98	−47.98	−47.99
3	−48.12	−48.12	−48.12	−48.11	−48.11	−48.11	−48.11	−48.12	−48.11	−48.11

3.1.4   Identify Optimal Level Values and Conduct Confirmation Experiment

Finding the largest S/N ratio in each column of Table 3.2 can identify the optimal level for that parameter. For example, the optimum levels in the first iteration are 2 for each parameter, as indicated by the italic emphasis in Table 3.2.

When the optimal levels are identified, a confirmation experiment is performed using the combination of the optimal levels identified in the response table. This confirmation test is not repetitious because the OA-based experiment is a fractional factorial experiment, and the optimal combination may not be included in Table 2.1. The fitness value obtained from the optimal combination is regarded as the fitness value of the current iteration.

3.1.5   Reduce the Optimization Range

If the results of the current iteration do not meet the termination criteria, which are discussed in the following subsection, the process is repeated in the next iteration. The optimal level values of the current iteration are used as central values (values of level 2) for the next iteration:

$$x_n\big|_{i+1}^2 = x_n\big|_i^{opt.}$$

$$(3.6)$$

To reduce the optimization range for a converged result, the LD_i is multiplied with a reduced rate (rr) to obtain LD_{i+1} for the $(i + 1)$th iteration:

$$LD_{i+1} = rr \cdot LD_i = rr^i \cdot LD_1 = RR(i) \cdot LD_1,$$

$$(3.7)$$

where $RR(i)$ is called *reduced function*. When a constant rr is used, $RR(i) = rr^i$. The value of rr can be set between 0.5 and 1 depending on the problem. The larger rr is, the slower the convergence rate. In this Rastigrin function optimization, rr is set to 0.8.

If LD_i is a large value, and the central level value $x_n|_i^2$ is located near the upper bound or lower bound of the optimization range, the corresponding value of level 1 or 3 may reside outside the optimization range. Therefore, a process of checking the level values is necessary to guarantee that all level values are located within the optimization range. If an excessive situation happens, reassigning the level value for the parameter will be performed. A simple way is to use the boundary values directly. For example, if $x_n|_i^1$ is smaller than min, the $x_n|_i^1$ is then set to min.

3.1.6   Check the Termination Criteria

When the number of iterations is large, the level difference of each element becomes small from (3.7). Hence, the level values are close to each other and the fitness value of the next iteration is close to the fitness value of the current iteration. The following equation may be used as a termination criterion for the optimization procedure:

$$\frac{LD_i}{LD_1} < \text{converged value}. \tag{3.8}$$

Usually, the *converged value* can be set between 0.001 and 0.01 depending on the problem. The iterative optimization process will be terminated *if the design goal is achieved* or *if (3.8) is satisfied.*

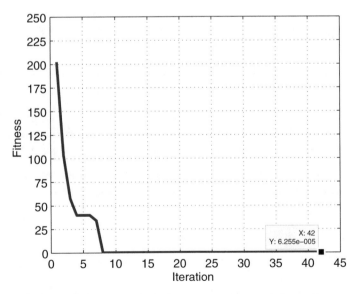

FIGURE 3.2: The fitness curve of (3.2). The optimized global minimum is 6.2553e-5 located at 1.453e-04, -1.193e-04, -7.738e-05, 2.456e-4, -1.156e-04, 2.212e-04, 8.147e-05, 2.882e-04, 2.483e-04, -8.899e-06.

In this Rastigrin function optimization, the optimization goal is that the fitness value shown in (3.2) should be smaller than 0.0001.

Following the aforementioned procedure, the optimization goal of (3.1) is obtained after 42 iterations. The convergence curve of the fitness function is shown in Figure 3.2. The optimized global minimum is 6.255e–5 when x_i, where $i = 1, \ldots, 10$ are 1.453e–04, –1.193e–04, –7.738e–05, 2.456e–4, –1.156e–04, 2.212e–04, 8.147e–05, 2.882e–04, 2.483e–04, and –8.899e–06.

3.2 OPTIMIZATION EXAMPLES

To evaluate the performance of an optimizer, it is important to choose the appropriate test functions. Pantoja et al. [34] suggested that test functions should contain properties with high-dimensional search spaces, multimodal functions, unimodal functions, and randomly perturbed functions. For example, (3.1) is a high-dimensional function and is used to investigate the optimizer's abilities for a problem with a large number of optimized variables.

3.2.1 Example 1

To demonstrate the global optimization performance of Taguchi's method, two relatively complicated two-dimensional test functions with many local maxima/minima are used. The first test function is similar to the equation presented in [35]:

$$\text{Fitness}(x,y) = \left| \frac{\sin\left[\pi\left(x-3\right)\right]}{\pi(x-3)} \right| \left\| \frac{\sin\left[\pi\left(y-3\right)\right]}{\pi(y-3)} \right| . \quad (0 \le x, y \le 8)$$

(3.9)

This function is useful to evaluate the convergence performance of an optimizer. The function has a global maximum, 1.0, located at ($x = 3.0, y = 3.0$) with many local maxima as shown in Fig. 3.3.

Taguchi's method is used for searching the global maximum of (3.9). The optimization range is 0 to 8 for both x and y directions. The reduced rate, rr, is set to 0.6; the *converged value* is set to 0.0001; and the OA(9, 2, 3, 2) is used in optimization. The optimized result is shown in Fig. 3.4. It can be seen that after only 11 iterations, the global maximum is successfully found by Taguchi's method. The optimized fitness is 1.00 located at ($x = 3.0037, y = 3.0037$).

3.2.2 Example 2

The second test function is a more challenging one, presented in [1], [36]:

$$\text{Fitness}(x,y) = x\sin(4x) + 1.1\,y\sin(2y). \quad (0 \le x, y \le 10)$$

(3.10)

The function is useful to detect anomalies of this kind [34]. It has a global minimum, –18.5547 located at ($x = 9.0390, y = 8.6682$) with many local minima close to the global minimum as shown in Fig. 3.5.

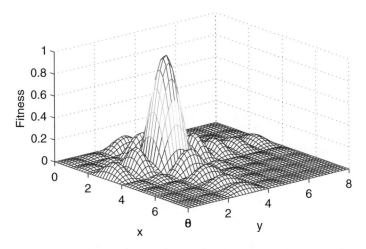

FIGURE 3.3: Three-dimensional solution surface of (3.9). The global maximum is 1.0 located at (x = 3.0, y = 3.0).

Again, Taguchi's method is used for searching the global minimum of (3.10). The optimization range is 0 to 10 for both *x* and *y* directions. The *converged rate* is set to 0.0001, and the OA(9, 2, 3, 2) is used in (3.10).

Different values of rr are used in the optimization process, and the corresponding optimization results are shown in Fig. 3.6. None of them can reach the global minimum. Optimization

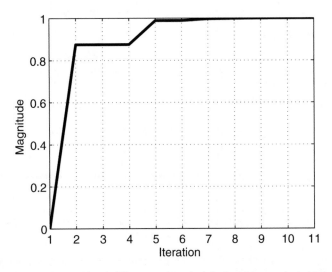

FIGURE 3.4: The fitness curve of (3.9). The optimized global maximum is 1.00 located at (x = 3.0037, y = 3.0037).

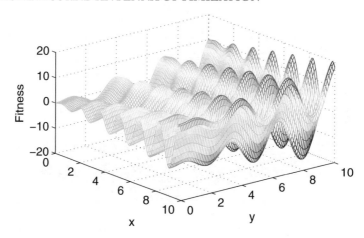

FIGURE 3.5: Three-dimensional solution surface of (3.10). The global minimum is −18.5547 located at $(x = 9.0390, y = 8.6682)$.

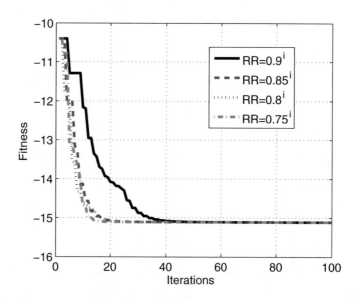

FIGURE 3.6: The fitness curve of (3.10). Different values of rr are used in the optimization process. It is noticed that the results are stuck at local optimums.

results are stuck at a local minimum because of the complexity in (3.10). The example shows that for sophisticated problems, the traditional Taguchi's method may not find the optimum. Therefore, special treatments and improved procedures for Taguchi's method are necessary to avoid ending up at a local optimum. These improved procedures are discussed in the following sections.

3.3   IMPROVED TAGUCHI'S METHOD

In the aforementioned implementation procedure, the optimization range is reduced as the iteration increases. When a constant rr is used, the optimization range becomes extremely small for the large iterations. Therefore, there is a possibility that optimized results may stick at a local optimum, as observed in Fig. 3.6. To improve the optimization performance of Taguchi's method, several techniques are suggested in this section.

3.3.1   Increase Initial Level Difference

LD_1 is determined by (3.3) and is reduced by rr or RR defined in (3.7) in each iteration. For the two-dimensional sophisticated problem like (3.10), the optimum solution is located near the edge of optimization range. The previous optimization process sticks at a local optimum since LD_i is too small to search the global optimum.

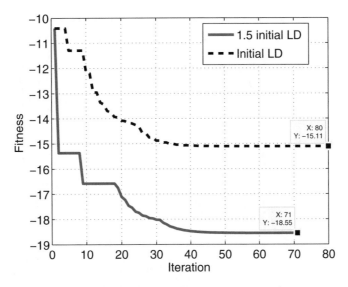

FIGURE 3.7: The optimized global minimum is −18.5546 located at ($x = 9.0401, y = 8.6694$) after 71 iterations. A total of 710 experiments were conducted. rr is set to 0.9.

The improved method introduced here is to increase the initial *level difference* (LD_1) by 1.5 times so that the optimization has a wider range to search the global optimum. The same test function (3.10) is used to prove the performance of this technique and the result is shown in Fig. 3.7. It is observed that the method can successfully find the global minimum after 71 iterations using the same three-level OA, with converged value = 0.0001 and rr set to 0.9.

Equation (3.9) is also used as an example to prove the performance of this method. Again, LD is increased 1.5 times. The result is shown in Fig. 3.8, which shows that the proposed technique can also successfully find the global maximum after 12 iterations using a three-level OA, OA(9, 2, 3, 2), with converged value = 0.0001 and rr set to 0.6. Since the initial level difference is increased, the number of iterations to reach the global optimum is also increased, which is a trade-off of the proposed technique.

3.3.2   Boundary Treatments

The current boundary condition treatment, shown in Section 3.1.5, mentioned that if $x_n|_i^1$ is smaller than min, $x_n|_i^1$ is then set to min. This strategy is simple and can work well in most problems. However, there is a possibility that two level values overlap at min or max together. This coincidence situation wastes the optimization energy. To overcome this drawback, an alternative method is proposed. When $x_n|_i^2$ (optimal level value) is located at min, in the next iteration:

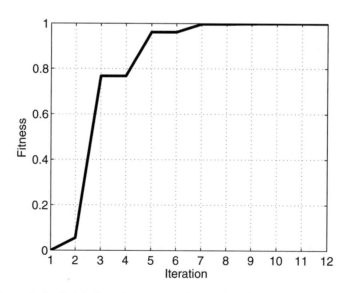

FIGURE 3.8: The optimized global maximum is 1.00 located at ($x = 2.9962, y = 2.9962$) after 12 iterations. A total of 120 experiments were conducted. rr is set to 0.6.

$$x_n\big|^1_{i+1} = \text{min};$$

$$(3.11)$$

$$x_n\big|^2_{i+1} = \text{min} + \text{LD(n)}_{i+1};$$

$$(3.12)$$

$$x_n\big|^3_{i+1} = \text{min} + 2 \times \text{LD}(n)_{i+1}.$$

$$(3.13)$$

When $x_n\big|^2_i$ is located at max, in the next iteration:

$$x_n\big|^3_{i+1} = \text{max};$$

$$(3.14)$$

$$x_n\big|^2_{i+1} = \text{max} - \text{LD}(n)_{i+1};$$

$$(3.15)$$

$$x_n\big|^1_{i+1} = \text{max} - 2 \times \text{LD}(n)_{i+1}.$$

$$(3.16)$$

Therefore, no two entries overlap at the same value. This method also guarantees that no entries are located out of the solution space, and the optimal level value is one of the entries.

3.3.3   Using a Gaussian Reduce Function

Another suggested technique is to set RR as a Gaussian function, i.e.,

$$\text{RR}(i) = e^{-\left(\frac{i}{T}\right)^2},$$

$$(3.17)$$

where T is the duration width of the Gaussian function. Values of T ranging from 15 to 20 are used in the examples presented here. Therefore, the *level difference* for the next iteration is:

$$\text{LD}_{i+1} = \text{RR}(i)\text{LD}_1 = e^{-\left(\frac{i}{T}\right)^2} \text{LD}_1.$$

$$(3.18)$$

The Gaussian function reduces the *level difference* slowly during the first several iterations to offer the optimization process more degrees of freedom to find the optimal solution while decreasing the *level difference* quickly to speed up the convergence of the optimization approach. A comparison of the Gaussian reduced function and the previous exponential reduced function is shown in Fig. 3.9.

The optimization performance is demonstrated by using (3.10). Optimized results are shown in Fig. 3.10, where two fitness curves for two reduced functions are plotted. LD is increased 1.5 times as discussed before. The converged value is set to 0.0001. After 42 iterations, the global minimum is obtained using the proposed Gaussian reduced function. In contrast, 71 iterations are required to reach the same solution with the traditional RR function, as shown in Fig. 3.10.

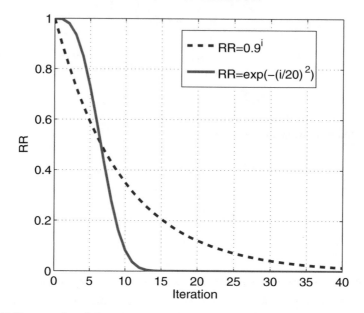

FIGURE 3.9: Different reduced functions versus iteration number.

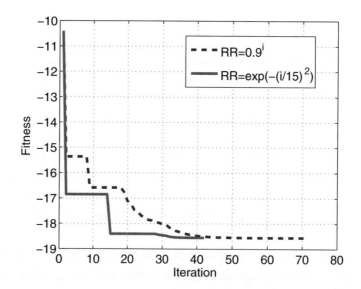

FIGURE 3.10: The optimized global minimum is −18.5546 located at ($x = 9.0379$, $y = 8.6673$), which was found after 42 iterations using the Gaussian function. The dotted line is the same as the solid one in Fig. 3.7.

Obviously, the optimization performance is improved by the Gaussian reduced function. It is quicker than the original RR function to find the optimum.

3.4 TAGUCHI'S METHOD WITH FIVE-LEVEL OA

In the previous examples, three-level OA are used in the optimization. OA with larger number of levels can also be used in Taguchi's optimization method. For example, five-level OA is used in this section. Instead of three possible options, now each parameter has five options in the optimization range. The strategy of using five-level OAs offers the optimization approach more power to search the global optimum. However, the higher the level of the OA, the larger the number of experiments should be conducted within each iteration.

Again, Taguchi's method is used for searching the global minimum of (3.10). The optimization range is 0 to 10 for both x and y directions. rr is set to 0.85; the *converged rate* is set to 0.0001. A five-level OA(25, 2, 5, 2), which is extracted from the first two columns in the OA(25, 6, 5, 2) shown in Table 2.3, is used in optimization. It is worthwhile to point out that the number of experiments per iteration is increased from 9 to 25 when the number of levels is increased from 3 to 5.

After 42 iterations, the global minimum is successfully found by Taguchi's method whereas using the three-level OA cannot reach the global minimum. The optimized result is shown in Fig. 3.11.

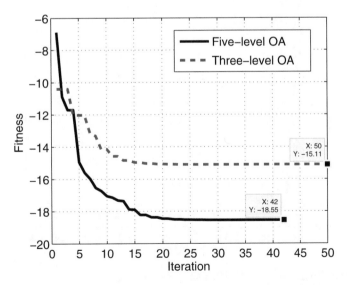

FIGURE 3.11: The fitness curve of (3.10) using a five-level OA. The optimized global minimum is −18.5546 located at ($x = 9.0399$, $y = 8.6673$). A total of 1092 experiments were conducted.

The optimized fitness is −18.5546 located at (x = 9.0399, y = 8.6673). To find the global minimum, a total of 1092 = (25 + 1) × 42 experiments were conducted.

3.5   RANDOM TAGUCHI'S METHOD

In the optimization process of Taguchi's method, the optimal levels are determined by a response table and the corresponding values are assigned to the central levels in the next iteration. Other entries are obtained by adding/subtracting the optimal level value with *level difference*, $LD(n)_i$. Thus, the search capability is confined by the level difference. When the level difference is small, the optimization may stick at local results.

To offer entries more optimization search capability, all level values except the central one are determined using random functions. This modified method is called here *Random-Taguchi's method*. For an optimization process of Random-Taguchi's method using a three-level OA, entries of level 1, $x_n|_i^1$, and level 3, $x_n|_i^3$, are determined by the following equations:

$$x_n|_{i+1}^1 = \text{min} + \text{rand}_1\,()\left[\,x_n|_{i+1}^2 - \text{min}\,\right], \tag{3.19}$$

$$x_n|_{i+1}^3 = x_n|_{i+1}^2 + \text{rand}_2\,()\left[\,\text{max} - x_n|_{i+1}^2\,\right], \tag{3.20}$$

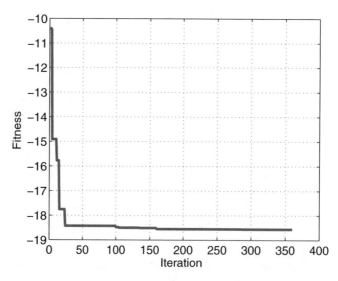

FIGURE 3.12: The fitness curve of (3.10) using Random-Taguchi's method. The optimized global minimum is −18.5547 located at (x = 9.0395, y = 8.6665) after 360 iterations using Random-Taguchi's method with three-level OA. A total of 3600 experiments were conducted.

where $\text{rand}_{1,2}()$ are uniform random values ranging from 0 to 1. Entries of $x_n|_i^1$ and $x_n|_i^3$ move freely between min and $x_n|_i^2$, and between $x_n|_i^2$ and max, respectively. Note that no rr or RR is required in the Random-Taguchi's method. Therefore, the proposed approach can avoid sticking at a local optimum.

Random-Taguchi's method is used for searching the global minimum of (3.10). The optimization range is 0 to 10 for both x and y directions; the *converged rate* is set to 0.0001; and a three-level OA, OA(9, 2, 3, 2), is used in (3.10). The optimized result is shown in Fig. 3.12. After 360 iterations, the global minimum is successfully found by Taguchi's method. The optimized fitness is −18.5547 located at ($x = 9.0395$, $y = 8.6665$).

From this example, it is clear that the Random-Taguchi's method can successfully avoid local minimum. However, one has to pay the price of the increased number of experiments. In this case, 3600 experiments were conducted to find the global minimum, which is more than that of traditional Taguchi's method. The reason is that random values are used in the optimization process.

•   •   •   •

CHAPTER 4

Linear Antenna Array Designs

4.1   INTRODUCTION OF LINEAR ANTENNA ARRAYS

Antenna array is an important area in EM and antenna engineering. It is used to realize specific radiation pattern, high antenna gain, and beam scanning capability. Antenna arrays are formed by assembling identical (in most cases) radiating elements such as dipoles and microstrip antennas. The array may be linear, planar, or volume array. In a linear array, the antenna elements are located along a straight line. A planar array has the elements distributed on a plane while a volume array has a three-dimensional distribution of the antenna elements. Linear antenna array is the basis for all antenna arrays and its design methodologies can be applied to other types of arrays.

Antenna pattern synthesis can be classified into several categories or groups [37, Chapter 7]. One of these groups requires that the antenna patterns possess nulls in desired directions. This property is widely used in smart antenna systems to eliminate the interference from specific noise directions. The Schelkunoff polynomial method is an effective approach to synthesize the null controlled patterns.

Another group requires that the antenna patterns exhibit a desired distribution in the entire visible region, which is also referred to as beam shaping. A typical example is the design of a sector beam pattern, which allows the antenna array to have a wider angular coverage. This is usually accomplished using the Fourier transform technique and the Woodward–Lawson method.

A third group usually requires antenna patterns with narrow beams and low side lobes. This guarantees the radiating or receiving energy to be more focused in specific directions. Various techniques such as the binomial method, Dolph–Chebyshev method, and Taylor line-source are proposed to serve this purpose.

In this chapter, Taguchi's optimization method is used to design three linear antenna arrays, where each array belongs to one of the groups mentioned above. Although they have different pattern requirements, the design process for the three arrays follows the same implementation procedure. It well demonstrates the versatility and robustness of Taguchi's optimization method.

4.2   ARRAY WITH NULL CONTROLLED PATTERN

The first design objective is to optimize the excitation magnitudes of array elements so that the corresponding array factor (AF) has nulls at specific directions [16–17, 19–21]. Fig. 4.1 depicts the

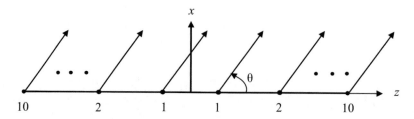

FIGURE 4.1: Geometry of a 20-element, equally spaced linear array.

antenna array geometry, which has 20 equally spaced elements along the z axis. The element spacing is a half-wavelength, and the excitations of the array elements are symmetric with respect to the x axis. Therefore, the excitation magnitudes of the 10 elements will be optimized in the range of [0, 1] to obtain an AF with prescribed nulls.

To use Taguchi's method, a proper OA needs to be selected first. Since there are 10 excitation magnitudes to be optimized, the OA to be selected should have 10 columns to represent these parameters. To characterize the nonlinear effect, three levels are necessary for each input parameter. Usually, an OA with a strength of 2 is efficient for most problems. Therefore, an OA with 10 columns, 3 levels, and 2 strengths is required. After searching the OA database [30], an OA(*27, 13, 3, 2*) is adopted for this problem. The first 10 columns of the OA are used for optimization process, and the remaining columns are ignored. Hence, the original OA becomes the OA(*27, 10, 3, 2*).

For a 20-element symmetrical array, the AF can be written as:

$$\text{AF}(\theta) = 2 \sum_{n=1}^{10} a(n) e^{j\varphi(n)} \cos[\,\beta d(n) \cos\theta\,], \tag{4.1}$$

where β is the wave number; $a(n)$, $\varphi(n)$, and $d(n)$ are the excitation magnitude, phase, and location of the nth element, respectively. Since only 10 excitation magnitudes should be optimized, the phase of each element is equal to 0 for this problem. The AF can be simplified as:

$$AF(\theta) = 2 \sum_{n=1}^{10} a(n) \cos[\,\beta\ d(n) \cos\theta\,]. \tag{4.2}$$

The fitness function is chosen according to the optimization goal. The following fitness function is used in the optimization:

$$\text{Fitness} = \int_{0°}^{180°} [\text{AF}(\theta) - \text{AF}_d(\theta)] \left[\frac{1 + \text{sgn}\,(\text{AF}(\theta) - \text{AF}_d(\theta))}{2} \right] d\theta, \tag{4.3}$$

where $AF(\theta)$ is the pattern obtained from (4.2), and it is a linear scale; $AF_d(\theta)$ is a mask for the desired null controlled pattern; and $d\theta$ is the angular interval, which is set to $0.1°$. The desired antenna pattern is shown in Fig. 4.2 by the dashed lines. Two nulls are desired to exist between $50°$ and $60°$, and between $120°$ and $130°$, having a magnitude of less than -55 dB. Basically, the fitness can be considered as the difference area between the desired pattern and the obtained pattern. When the obtained pattern is within the mask, $AF(\theta) < AF_d(\theta)$, its contribution to fitness is 0. When the obtained pattern is outside the mask, $AF(\theta) > AF_d(\theta)$, it increases the fitness value. Apparently, the smaller the fitness value, the better the match between the obtained pattern and the desired one.

In this optimization case, the converged value is set to 0.002, and rr is set to 0.75. When this optimization process has been executed for 23 iterations, an optimal null control pattern is obtained, as presented in Fig. 4.2. The result shows that the beam width at -40 dB side lobe level (SLL) is $20.9°$, the half power beam width (HPBW) is $7.4°$, and nulls are below -55 dB in the angle ranges of $[50°, 60°]$ and $[120°, 130°]$, as desired. The optimized excitation magnitudes of the elements from number 1 to number 10 are as follows: [0.603, 0.577, 0.530, 0.459, 0.379, 0.294, 0.214, 0.155, 0.082, 0.046], as shown in Fig. 4.3. To illustrate the efficiency of Taguchi's method, the convergence

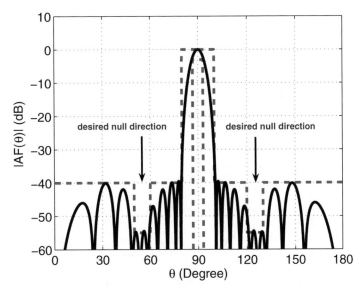

FIGURE 4.2: Null controlled pattern of an optimized 20-element linear array. The dashed lines are the desired pattern, which has prescribed nulls at $[50°, 60°]$ and $[120°, 130°]$ with level of -55 dB and has a 3-dB beam width of $7.4°$. From [17], copyright © IEEE 2007.

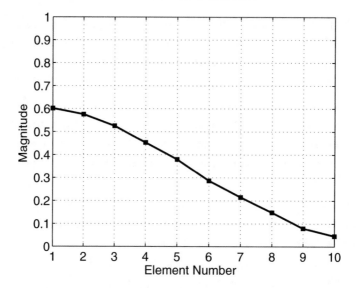

FIGURE 4.3: Optimized excitation magnitudes of the linear antenna array with a null-controlled pattern shown in Fig. 4.2. From [17], copyright © IEEE 2007.

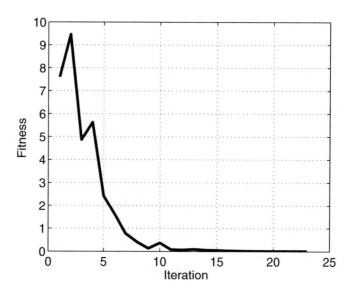

FIGURE 4.4: Convergence curve of the fitness value of the 20-element, equally spaced linear array for the null-controlled pattern design. From [17], copyright © IEEE 2007.

curve of the fitness value is plotted in Fig. 4.4. It is observed that the fitness value converges to the optimum result quickly.

4.3   ARRAY WITH SECTOR BEAM PATTERN

To further demonstrate the validity of Taguchi's method, a relatively complex case, a sector beam pattern design, is attempted here. In this design, the same 20-element array in Fig 4.1 is used again, but both excitation magnitudes and phases of the array elements are to be optimized to shape the antenna pattern [23], [38, Chapter 3]. Thus, an OA(*81, 20, 3, 2*) [30], which offers 10 columns of magnitudes and 10 columns of phases, is adopted in this sector beam pattern synthesis.

The requirements for the sector beam pattern are shown in Fig. 4.5 using dashed lines. To define the sector beam, there are two specific angular regions. Region I ranges from 78° to 102°, where ripples should be smaller than 0.5 dB. Region II controls SLLs, which are all below −25 dB between 0° and 70° and between 110° and 180°.

The optimization ranges of the excitation magnitude and phase of each element are from 0 to 1 and from −π to π, respectively. Equation (4.3) is used for evaluating the fitness value during the optimization process. Since this sector beam pattern problem is more sophisticated, rr may be

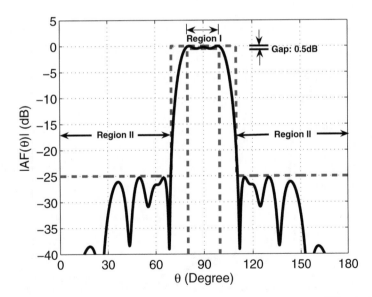

FIGURE 4.5: Sector beam pattern of an optimized 20-element linear array. The dashed lines are the desired pattern mask, which requires ripples in region I smaller than 0.5 dB and SLLs in region II lower than −25 dB. From [17], copyright © IEEE 2007.

set to a larger value to offer fine searching ability to the optimizer. rr is set to 0.9 in this case. The converged value is set to 0.002.

The convergence curve of the fitness value is presented in Fig. 4.6. After 60 iterations, an optimum sector beam pattern is obtained as plotted in Fig. 4.5. It has a ripple of 0.48 dB in region I, the beam width at −25 dB SLL is 41.2°, and the HPBW is 28.1°. The optimized excitation magnitudes of the elements from number 1 to number 10 are [0.437, 0.321, 0.188, 0.122, 0.132, 0.130, 0.079, 0, 0, 0], as shown in Fig. 4.7(a). The optimized excitation phases (degree) of the elements are [9.03, 2.51, −16.74, −77.72, −119.81, −112.63, −111.57, −111.27, −170.14, −175.43], as shown in Fig. 4.7(b). The optimized result indicates that a 14-element symmetrical linear array is capable of realizing the same design goal but with less antenna elements.

The same sector beam problem was optimized in [23] using the PSO method. The desired pattern was obtained by using 20 particles and running around 800 iterations. Thus, the total experiments needed were 16 000. However, only 4920 experiments (82 experiments by 60 iterations) are required to achieve the same goal by Taguchi's method. A comparison plot is presented in Fig. 4.8. The reduction in the number of experiments is around 70%, which shows that Taguchi's method is quicker than PSO to achieve the same optimization goal for this problem.

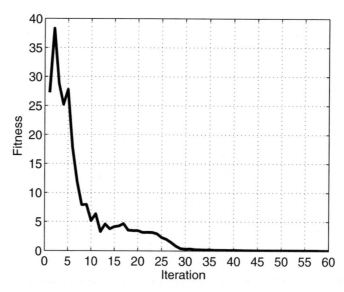

FIGURE 4.6: Convergence curve of the fitness value for the 20-element, equally spaced linear array of the sector beam pattern design. From [17], copyright © IEEE 2007.

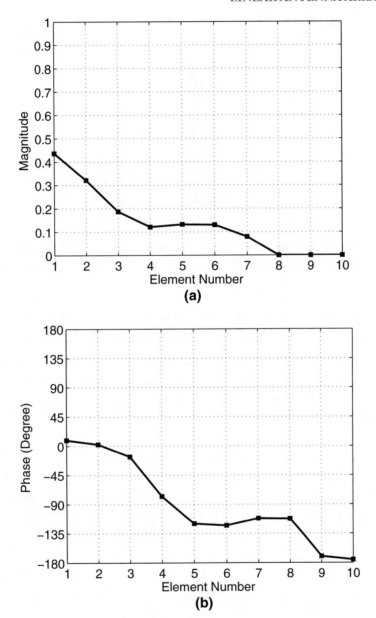

FIGURE 4.7: Optimized element excitations of the linear antenna array with a sector beam pattern shown in Fig. 4.5. (a) The magnitudes of elements and (b) the phases of elements. From [17], copyright © IEEE 2007.

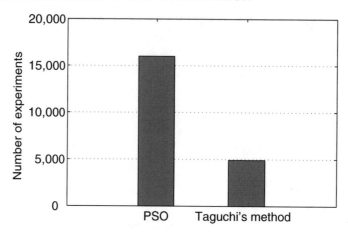

FIGURE 4.8: Comparison of total numbers of experiments required in PSO [22] and in Taguchi's optimization method for the sector beam pattern optimization problem. From [17], copyright © IEEE 2007.

4.4   ARRAY WITH SUPPRESSED SIDE LOBE LEVELS

Taguchi's method is also applied in the optimization of an unequally spaced linear array [19, 24–25]. The geometry of a symmetric 10-element unequally spaced linear array is shown in Fig. 4.9. The element positions are located within a given aperture size of 5 λ. The excitation of each element is uniform in this case, and the locations of the elements are optimized to suppress the SLL. Therefore, the AF for this case is written as

$$AF(\theta) = 2\sum_{n=1}^{5} \cos[kd(n)\cos(\theta)]. \qquad (4.4)$$

The goal of this optimization is to suppress the SLL under −18.96 dB, which is reduced by 6 dB compared with that of an equally spaced linear array [25]. Meanwhile, it is required to maintain

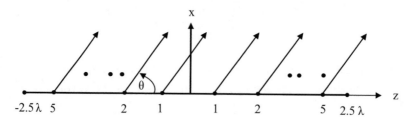

FIGURE 4.9: Geometry of a 10-element, unequally spaced linear array. The element excitations are uniform, and the locations are optimized to suppress the SLLs.

TABLE 4.1: An OA(18, 5, 3, 2) is used for the unequally spaced linear array design

EXPERIMENTS	ELEMENTS				
	1	2	3	4	5
1	1	1	1	1	1
2	2	2	2	2	2
3	3	3	3	3	3
4	1	1	2	3	2
5	2	2	3	1	3
6	3	3	1	2	1
7	1	2	1	3	3
8	2	3	2	1	1
9	3	1	3	2	2
10	1	3	3	1	2
11	2	1	1	2	3
12	3	2	2	3	1
13	1	2	3	2	1
14	2	3	1	3	2
15	3	1	2	1	3
16	1	3	2	2	3
17	2	1	3	3	1
18	3	2	1	1	2

TABLE 4.2: Initial level values of each element in the first iteration of a 10-element, unequally spaced linear array problem (unit: wavelength)

EXPERIMENTS	ELEMENTS				
	1	2	3	4·	5
1	0	0.5	1	1.5	2
2	0.25	0.75	1.25	1.75	2.25
3	0.5	1	1.5	2	2.5
4	0	0.5	1.25	2	2.25
5	0.25	0.75	1.5	1.5	2.5
6	0.5	1	1	1.75	2
7	0	0.75	1	2	2.5
8	0.25	1	1.25	1.5	2
9	0.5	0.5	1.5	1.75	2.25
10	0	1	1.5	1.5	2.25
11	0.25	0.5	1	1.75	2.5
12	0.5	0.75	1.25	2	2
13	0	0.75	1.5	1.75	2
14	0.25	1	1	2	2.25
15	0.5	0.5	1.25	1.5	2.5
16	0	1	1.25	1.75	2.5
17	0.25	0.5	1.5	2	2
18	0.5	0.75	1	1.5	2.25

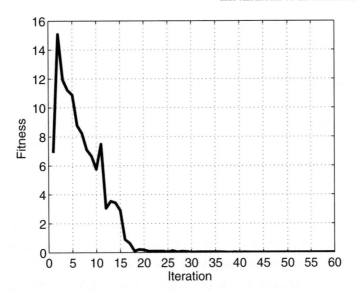

FIGURE 4.10: Convergence curve of the fitness value of the 10-element, unequally spaced linear array for the suppressed SLL design.

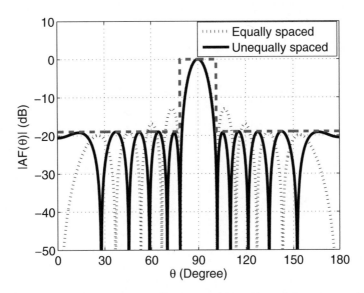

FIGURE 4.11: Array factor of the 10-element linear array. The SLL of the optimized unequally spaced array is 6 dB lower than that of the equally spaced array.

a similar beam width. Equation (4.3) is used again for evaluating the fitness value during the optimization process.

An OA(*18, 5, 3, 2*) [30] shown in Table 4.1, which offers five optimization parameters corresponding to the location of five antenna elements, is used in this study. To start the optimization-process, the initial level values of the parameters $d(n)\big|_1^2$ are set to the locations of an equally spaced linear array, and LD_1 is set to quarter wavelength. Table 4.2 shows the initial values of each element in the first iteration. In this optimization case, the converged value and rr are set to 0.002 and 0.9, respectively.

After 60 iterations, an optimized pattern is obtained since the fitness is converged, as shown in Fig. 4.10. The pattern of the unequally spaced array is compared with the equally spaced array pattern in Fig. 4.11. It is observed that the SLL is successfully suppressed to −18.96 dB while maintaining the same main beam shape as that of the equally spaced array. In the optimized result, the side lobes have approximately the same level. The beam width is 20.7°, SLL is −18.96 dB, and HPBW is 9.8°. The optimized locations of the elements from number 1 to number 5 are [0.1996 λ, 0.6588 λ, 1.1166 λ, 1.7190 λ, 2.4195 λ].

•   •   •   •

CHAPTER 5

Planar Filter Designs

5.1 INTRODUCTION OF PLANAR FILTERS

Microwave filters are widely used in telecommunication equipments. Filters are used to suppress the noises coming from the environment, prevent spurious signals from interfering with other systems, and allow desired signals to pass through within a specific frequency band. Although filters built with lumped elements can realize the desired frequency responses, it is difficult to control the lumped elements' properties in the microwave band. Instead, filters consisting of distributed transmission line elements are usually used in microwave applications.

The printed planar-type filters are composed of several stubs of microstrip lines or coplanar strips, and need to be accurately modeled due to high-frequency effects such as dispersion and dielectric/conductor loss. Therefore, *an efficient optimization technique* and *a full-wave EM simulator* are necessary tools for an optimum design of such filters. In this study, a full-wave commercial software (IE3D [26]) along with an external Taguchi's method-based optimizer is used to design various filters, namely, a microstrip band stop filter (BSF), a coplanar waveguide BSF, and a microstrip band pass filter (BPF). The desired frequency responses of the designed filters are successfully achieved with only a few numbers of iterations.

5.2 INTEGRATION OF TAGUCHI'S METHOD WITH IE3D

Fig. 5.1 shows the flowchart of the optimization procedure. The initial dimensions of a filter are used as starting points in the optimization process. IE3D, a full-wave EM simulator based on method of moments, is utilized to compute S parameters of the filter. In IE3D, an EM problem is analyzed according to *.sim and *.geo input files, which contain the simulation information and the dimensions of optimized parameters. The simulation results are stored in an *.sp output file.

Taguchi's method is applied as the external optimizer to drive the IE3D engine. The optimization code is developed such that it can change the contents of the ..sim file to control the IE3D simulation. Fig. 5.2 shows the partial contents of an *.sim file. In each iteration, the dimensions of a filter, as indicated by a grey background in Fig. 5.2, are determined by Taguchi's method. Note that the value of a dimension is an offset value from the initial dimension of each parameter, and it may not exceed either the higher bound or the lower bound of the parameter.

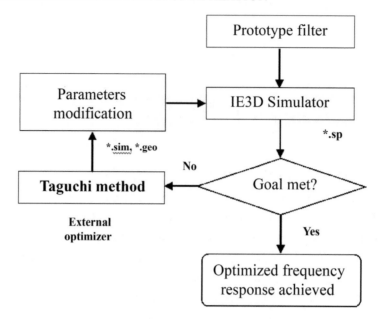

FIGURE 5.1: Flowchart of filter optimization.

```
<OptimVariables id="opt" name_convention="1" >
<ObjList id="this">
<OptimVariable comment="L5" End="0" HighBound="1.5" LowBound="-1.5" Noffset="1" Start="-0.523"
StepDeriv="0" >
</OptimVariable>
<OptimVariable comment="L4" End="0" HighBound="1" LowBound="-1" Noffset="1" Start="0.410"
StepDeriv="0" >
</OptimVariable>
<OptimVariable comment="S1" End="0" HighBound="0.05" LowBound="-0.05" Noffset="1" Start="0.009"
StepDeriv="0" >
</OptimVariable>
<OptimVariable comment="S2" End="0" HighBound="0.1" LowBound="-0.1" Noffset="1" Start="-0.037"
StepDeriv="0" >
</OptimVariable>
<OptimVariable comment="S3" End="0" HighBound="0.1" LowBound="-0.1" Noffset="1" Start="-0.054"
StepDeriv="0" >
</OptimVariable>
<OptimVariable comment="L1" End="0" HighBound="0.5" LowBound="-0.5" Noffset="1" Start="0.164"
StepDeriv="0" >
</OptimVariable>
<OptimVariable comment="L2" End="0" HighBound="0.5" LowBound="-0.5" Noffset="1" Start="0.252"
StepDeriv="0" >
</OptimVariable>
<OptimVariable comment="L3" End="0" HighBound="0.5" LowBound="-0.5" Noffset="1" Start="0.000"
StepDeriv="0" >
</OptimVariable>
</ObjList>
</OptimVariables>
```

FIGURE 5.2: Partial contents of an *.sim file.

After each simulation experiment, the optimization code can read the simulated S parameter results from the *.sp file to calculate the fitness value. Fig. 5.3 shows the partial contents of an *.sp file. The first five rows contain the output message. The first column contains frequency points, which are set by the user. The second column contains the magnitudes of S_{11} in linear scale. The third column represents the phases of S_{11} in degrees. If the number of columns is more than 3, the remaining columns are other S parameters by orders of S_{12}, S_{21}, and S_{22}, respectively.

Once the full-wave simulations of all experiments in the same iteration are finished, the computed fitness values are used to build a response table and optimum levels are identified accordingly. The remaining procedure is the same as the Taguchi method discussed in Chapter 3. If the results

```
! Zeland S-Parameters Output Version 2.0
# GHZ S MA R 50
! Nport = 2
!

5            0.32827      -167.78      0.93846      -77.115
5.15         0.32856      -170.16      0.93839      -79.545
5.3          0.32792      -172.58      0.93858      -82.003
5.45         0.3263       -175.04      0.93905      -84.493
5.6          0.32361      -177.53      0.93984      -87.021
5.75         0.31975      179.92       0.94096      -89.593
5.9          0.31463      177.33       0.94243      -92.215
6.05         0.30812      174.68       0.94429      -94.894
6.2          0.30009      171.97       0.94653      -97.639
6.35         0.2904       169.2        0.94916      -100.46
6.5          0.27886      166.36       0.95219      -103.36
6.65         0.26527      163.44       0.9556       -106.37
6.8          0.2494       160.44       0.95935      -109.48
6.95         0.23101      157.37       0.96339      -112.72
7.1          0.20979      154.23       0.96764      -116.1
```

FIGURE 5.3: Partial contents of an *.sp file.

do not meet the termination criteria, the dimensions of the filter in *.sim file are modified for the next iteration. The procedure described by the flowchart in Fig. 5.1 is repeated until the optimum result is obtained.

5.3   MICROSTRIP BAND STOP FILTER

A symmetrical double-folded stub microstrip BSF [39, 40], as shown in Fig. 5.4, is used as the first example to demonstrate the validity of Taguchi's method in filter design. The thickness of the substrate is 5 mil, and the relative dielectric constant is 9.9. The characteristic impedance of all microstrip lines is 50 Ω, and the corresponding width W is 4.8 mil.

The filter specifications, which are drawn as dotted lines ($\left|S_{21,d}\right|$) in Fig. 5.5, are defined as

Region I:
$$\left|S_{21,d}\right| < -30\,\text{dB} \quad \text{for } 12\,\text{GHz} < f < 14\,\text{GHz} \tag{5.1}$$

Region II:
$$\left|S_{21,d}\right| < -3\text{dB} \quad \text{for } 16.5\,\text{GHz} < f \text{ and } f < 9.5\,\text{GHz} \tag{5.2}$$

where $\left|S_{21,d}\right|$ is the magnitude of the desired transmission coefficient. In Fig. 5.4, three parameters, L_1, L_2, and S, are to be optimized to achieve the design goal of the BSF. An OA(*9, 3, 3, 2*) [30], which offers three columns for L_1, L_2, and S, is adopted in this BSF optimization and is shown in Table 5.1. At the beginning of the optimization process, the initial filter dimension of L_1 is 74 mil, L_2 is 62 mil, and S is 13 mil. The optimization ranges of L_1, L_2, and S are from 54 to 94, 32 to 92, and 4 to 21 mil, respectively.

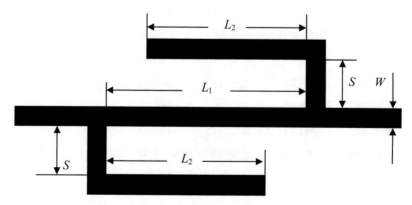

FIGURE 5.4: Geometry of the microstrip band-stop filter.

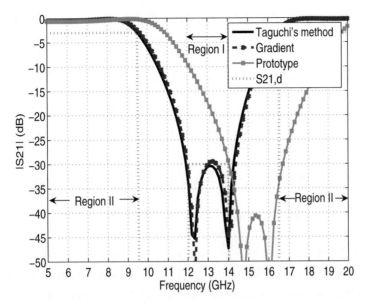

FIGURE 5.5: The optimized $|S_{21}|$ compared with the $|S_{21}|$ obtained by a gradient method, the $|S_{21}|$ of the original prototype BSF, and the design specification ($|S_{21,d}|$).

	L_1	L_2	S
TABLE 5.1: A three-level OA(9, 3, 3, 2) for three parameters L_1, L_2, and S			
1	1	1	1
2	1	2	3
3	1	3	2
4	2	2	2
5	2	3	1
6	2	1	3
7	3	3	3
8	3	1	2
9	3	2	1

The following fitness function is used in the optimization:

$$Fitness = w_1 \left\{ \int_{12}^{14} \left(\left| S_{21} \right| - \left| S_{21,\,d} \right| \right)_{dB} \left[\frac{1 + \text{sgn}\left(\left| S_{21} \right| - \left| S_{21,\,d} \right| \right)}{2} \right] df \right\}$$

$$+ w_2 \left\{ \int_{5}^{9.5} \left(\left| S_{21,d} \right| - \left| S_{21} \right| \right)_{dB} \left[\frac{1 + \text{sgn}\left(\left| S_{21,d} \right| - \left| S_{21} \right| \right)}{2} \right] df \right\}$$

$$+ \int_{16.5}^{20} \left(\left| S_{21,d} \right| - \left| S_{21} \right| \right)_{dB} \left[\frac{1 + \text{sgn}\left(\left| S_{21,d} \right| - \left| S_{21} \right| \right)}{2} \right] df \right\}, \qquad (5.3)$$

where w_1 and w_2 are the weights of region I fitness and region II fitness, respectively; the unit of frequency is gigahertz; and the df is the frequency interval set to 0.15 GHz. Basically, the fitness can be seen as the difference area which the obtained $|S_{21}|$ exceeds the desired $|S_{21,d}|$ in region I, and the desired $|S_{21,d}|$ exceeds the obtained $|S_{21}|$ in region II. The smaller the value of the fitness function is, the closer the results are toward the desired design goal. The ideal fitness value is 0, which means that the optimized results fully satisfy the design goal. Since $|S_{21}|$ (decibel scale) in region II is smaller than that in region I, to balance the priority in the two regions, the w_2 should be set larger than w_1. Therefore, in this BSF optimization, the value of w_1 is set to 1, and w_2 is set to 5. In addition, the converged value is set to 0.05, and rr is set to 0.8.

After 11 iterations, the fitness value reaches 0, and the optimization process ends due to the fact that the design goal was achieved. The convergence curve of fitness is presented in Fig. 5.6,

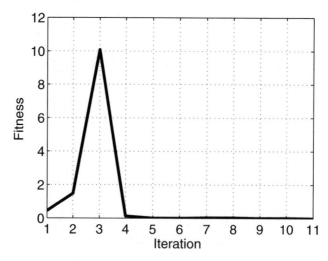

FIGURE 5.6: Convergence curve of the fitness value of the BSF design.

PARAMETERS	PROTOTYPE	OPTIMIZED BY GRADIENT [39]	OPTIMIZED BY TAGUCHI'S METHOD
L_1	74.0	91.82	92.26
L_2	62.0	84.71	84.64
S	13.0	4.80	5.12

TABLE 5.2: The dimensions (in mils) of prototype BSF, obtained by Taguchi's method, and optimized by the gradient method

which demonstrates the efficiency of Taguchi's method. The dimensions of the initial BSF, obtained by Taguchi's method, and the ones optimized by the gradient method [39] are listed in Table 5.2.

The $|S_{21}|$ of the BSF optimized by Taguchi's method, the $|S_{21}|$ of the BSF optimized by the gradient method [39], the $|S_{21}|$ of the original prototype BSF, and the design specification $|S_{21,d}|$ are all shown in Fig. 5.5. The desired frequency response of S_{21} is achieved, which demonstrates the validity of Taguchi's method.

5.4   COPLANAR WAVEGUIDE BAND STOP FILTER

A coplanar waveguide (CPW) line with a compact BSF [41], as shown in Fig. 5.7, is used as the second example to demonstrate the validity of Taguchi's method in filter design. The BSF is fabricated on the *Rogers RO 4003C* substrate with the thickness of the substrate as 0.813 mm. The relative dielectric constant is 3.38. The characteristic impedance of the CPW line is 50 Ω. The corresponding width W_f is 5.0 mm, and the gap g between ground plane and the CPW line is 0.25 mm.

A stop-band feature at the center frequency (5.5 GHz) of the WLAN frequency band (5.15–5.825 GHz) is desired. The design specification is taken as:

$$\left|S_{21}\right| < -35 \text{ dB   at   5.5 GHz}. \tag{5.4}$$

In Fig. 5.7, four parameters, W_1, L_1, W_2, and L_2, are optimized to achieve the design goal of the BSF. An OA(*9, 4, 3, 2*) [30], which offers four columns for W_1, L_1, W_2, and L_2, is adopted in this BSF optimization. The dimensions and optimization ranges of the initial BSF are shown in Table 5.3.

The following fitness function is used in the optimization:

$$Fitness = \left\|S_{11}\left(f = 5.5\,GHz\right)\right\|. \tag{5.5}$$

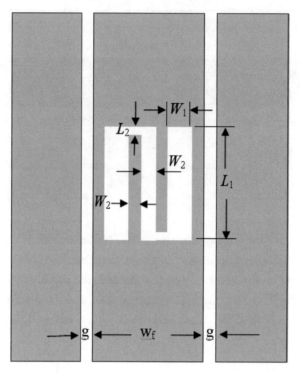

FIGURE 5.7: Geometry of the CPW line with a compact band-stop filter.

Note that the fitness value is the absolute value of S_{11} in decibel scale. The smaller the value of the fitness function is, the better the band-stop performance at 5.5 GHz. The ideal fitness value is 0, which means that the incoming signal is completely rejected by the BSF. The fitness value is converted to the S/N ratio (η) in Taguchi's method using (3.5) to build the response table. Hence, a small fitness value results in a large S/N ratio. The response table is created in each iteration. For example, the response table of the first iteration is shown in Table 5.4.

Not only can the response table identify the optimal combination of parameters, but it can also determine the parameters' importance in affecting the filter performance. The difference between the maximum S/N ratio and the minimum S/N ratio of a parameter indicates the relative influence of the output, which is the fitness or S parameter response in this case. The larger the difference is, the stronger the influence. In the first iteration, the differences of S/N ratios of four parameters are calculated and also shown in Table 5.4. Therefore, the largest difference is ranked number one for L_1, which means that L_1 is the most influential parameter affecting fitness in the first iteration.

TABLE 5.3: The initial values, optimization ranges, and optimized values of the BSF parameters (all dimensions are in millimeters).

PARAMETERS	W_1	L_1	W_2	L_2
Initial	0.75	4.75	0.4	0.4
Optimization range	0.75–1.29	3.95–6.35	0.4–0.66	0.2–0.5
Optimized	1.29	5.60	0.65	0.38

In this BSF optimization case, the converged value is set to 0.01, and rr is set to 0.8. Only after six iterations does the optimization process end due to the fact that the design specification in (5.4) is achieved. The fitness value at the sixth iteration is 0.125. The convergence curve of fitness is presented in Fig. 5.8, which again demonstrates the efficiency of Taguchi's method.

The optimized dimensions of the optimized BSF are listed in Table 5.3 and are used to design a UWB antenna with a band-stop feature in the next chapter. Before the optimization process, the $|S_{11}|$ and $|S_{21}|$ of the initial BSF are quite poor as shown in Fig. 5.9. The $|S_{11}|$ and $|S_{21}|$ of the BSF optimized by Taguchi's method are also shown in Fig. 5.9 for comparison. The frequency response $|S_{21}|$ of the optimized BSF at 5.5 GHz is −38 dB, which satisfies the optimization goal.

TABLE 5.4: Response table after the first iteration of the CPW BSF optimization (decibel)

η	PARAMETERS			
	W_1	L_1	W_2	L_2
Level 1	−22.948	−23.407	−21.989	−20.843
Level 2	−21.49	−22.257	−20.983	−21.89
Level 3	−19.574	−18.347	−21.039	−21.279
$|\eta_{max} - \eta_{min}|$	3.374	5.060	1.006	1.048
Ranking	2	1	4	3

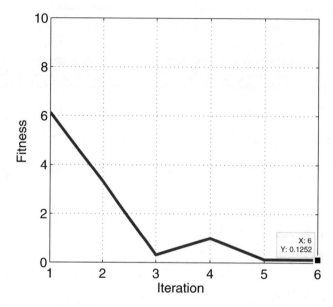

FIGURE 5.8: Convergence curve of the fitness value of the CPW BSF design.

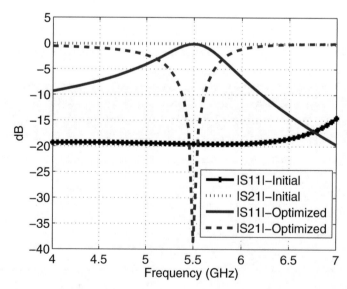

FIGURE 5.9: The $|S_{11}|$ and $|S_{21}|$ of the initial BSF and optimized BSF.

ITERATION	PARAMETERS			
	W_1	L_1	W_2	L_2
1	2	1	4	3
2	4	1	2	3
3	4	1	2	3
4	3	1	2	4
5	3	1	2	4
6	3	1	2	4
Mean of ranking	3.17	1.00	2.33	3.50
Overall ranking	3	1	2	4

TABLE 5.5: Rankings of the CPW BSF optimization parameters

To determine which parameter has the major effect, the rankings of the four parameters are recorded in each iteration, as shown in Table 5.5. The mean of ranking is calculated for each parameter, and the overall ranking is identified as well. Therefore, the most influential parameter affecting the fitness or input impedance matching of the BSF is L_1. This ranking provides good guidance to achieve the proper tolerances while manufacturing the filter. The capability of identifying the most important design parameters is another useful feature of the proposed Taguchi's method.

5.5   MICROSTRIP BAND PASS FILTER

A six-pole, edge-coupled, microstrip BPF is shown in Fig. 5.10 as the third optimization example using Taguchi's method [42]. In the geometry of the proposed filter, the width (W) of the coupled lines (L_1 to L_3) and resonant lines (L_4 and L_5) is fixed to 0.26 mm. The width (W_{50}) and length (L_{50}) of the 50-Ω microstrip lines, which connect the input and output ports, are also fixed to 0.567 and 10.0 mm, respectively. The BPF is designed on a substrate with a thickness of 0.64 mm, a dielectric constant of 10.2, and a dielectric loss tangent of 0.0023 according to the manufacturer specifications of *Rogers RO 3010* material.

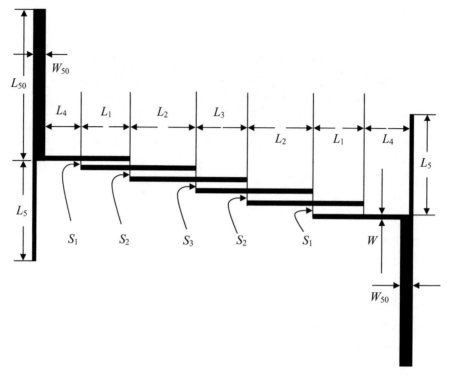

FIGURE 5.10: Geometry of the six-pole, edge-coupled, microstrip band-pass filter. From [42], copyright © IEEE 2007.

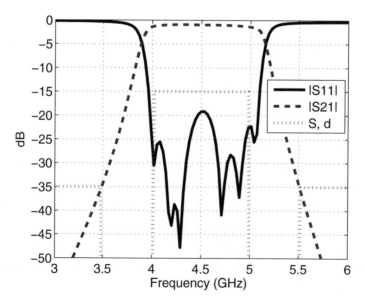

FIGURE 5.11: Optimized S parameters of the band-pass filter.

The filter specifications, which are drawn as dotted lines in Fig. 5.11, are taken from [43]:

Region I: $\qquad\qquad |S_{11}| < -15\,\mathrm{dB},\quad \text{for } 4\,\mathrm{GHz} < f < 5\,\mathrm{GHz},$ (5.6)

Region II: $\qquad\quad |S_{21}| < -35\,\mathrm{dB},\quad \text{for } f < 3.5\,\mathrm{GHz} \text{ and } f > 5.5\,\mathrm{GHz}.$ (5.7)

The same optimization procedure described in the previous section is applied again. The dimensions of the BPF are modified in each iteration until the results of the S parameters meet the termination criteria.

The initial length of coupled lines (L_1–L_3) and resonant lines (L_4, L_5) is set to 6.2 mm, which is around a quarter wavelength at 4.5 GHz. The initial spacing between the coupled lines (S_1–S_3) can be obtained by the parallel-coupled line formula$s$ [43]. Therefore, the initial values of S_1, S_2, and S_3 are equal to 0.254, 0.413, and 0.432 mm, respectively. A summary of the initial dimensions of the BPF is shown in Table 5.6.

TABLE 5.6: The initial dimensions, optimization ranges, and optimized dimensions obtained by Taguchi's method of the BPF case shown in Fig. 5.10 (all dimension are in millimeters)

PARAMETERS	INITIAL DIMENSIONS	OPTIMIZATION RANGE	OPTIMIZED DIMENSIONS
L_1	6.2	5.7–6.7	6.364
L_2	6.2	5.7–6.7	6.452
L_3	6.2	5.2–7.2	6.200
L_4	2.1	1.1–3.1	1.690
L_5	4.1	2.6–5.6	3.577
S_1	0.254	0.204–0.304	0.263
S_2	0.413	0.313–0.513	0.376
S_3	0.432	0.332–0.532	0.378
L_{50}	10.0	0	10.0
W_{50}	0.567	0	0.567
W	0.26	0	0.26

Since there are eight parameters (L_1–L_5; S_1–S_3) that need to be optimized to achieve the design specifications of the BPF, an OA(*27, 8, 3, 2*) [30] that offers eight columns is adopted in the BPF optimization. The selected optimization ranges of the eight parameters are also shown in Table 5.6. Note that the optimization range of S_1 is smaller than that of S_2 and S_3 to restrict the narrowest spacing between coupled lines to 0.2 mm.

The following fitness function is used in the optimization process:

$$
\begin{aligned}
\text{Fitness} = w_1 & \left\{ \int_{4.0}^{5.0} \left(\left| S_{11} \right| - \left| S_{11,\,d} \right| \right)_{dB} \left[\frac{1 + \text{sgn}\left(\left| S_{11} \right| - \left| S_{11,\,d} \right| \right)}{2} \right] df \right\} \\
+ w_2 & \left\{ \int_{3.0}^{3.5} \left(\left| S_{21} \right| - \left| S_{21,d} \right| \right)_{dB} \left[\frac{1 + \text{sgn}\left(\left| S_{21} \right| - \left| S_{21,d} \right| \right)}{2} \right] df \right\} \\
+ & \left. \int_{5.5}^{6.0} \left(\left| S_{21} \right| - \left| S_{21,d} \right| \right)_{dB} \left[\frac{1 + \text{sgn}\left(\left| S_{21} \right| - \left| S_{21,d} \right| \right)}{2} \right] df \right\} \quad ,
\end{aligned}
\tag{5.8}
$$

where w_1 and w_2 are identical in this example. The unit of frequency is gigahertz, and the df is the frequency interval which is set to 0.03 GHz. The fitness value can be seen as the difference area

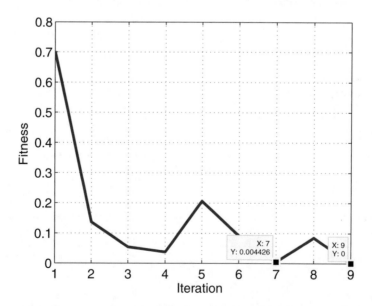

FIGURE 5.12: Convergence curve of the fitness value of the BPF design.

where the obtained $|S_{21}|$ exceeds the desired $|S_{21,d}|$ in region II, and the obtained $|S_{11}|$ exceeds the desired $|S_{11,d}|$ in region I. Therefore, a smaller fitness value reflects a better filter design.

In the Taguchi's optimization process, the *converged rate* is set to 0.05, and rr is set to 0.8. After nine iterations, the fitness value approaches 0 and the optimization process ends. The convergence curve of fitness is presented in Fig. 5.12. The optimized dimensions are listed in Table 5.6. The $|S_{11}|$ and $|S_{21}|$ of the optimized BPF are shown in Fig. 5.11. The desired frequency responses of S parameters are achieved, which again demonstrates the validity and efficiency of Taguchi's method.

•   •   •   •

CHAPTER 6

Ultra-Wide Band (UWB) Antenna Designs

6.1 INTRODUCTION OF UWB ANTENNA

In 2002, the Federal Communications Commission (FCC) released the ultra-wideband (UWB) system whose spectrum covers from 3.1 to 10.6 GHz [27]. Since then, UWB antennas have attracted more and more attention in the EM community. Recently, many UWB antennas have been designed and published. Some of them are nonplanar antennas [44–46] whose ground planes are vertical to the radiators as shown in Fig. 6.1. In contrast, planar UWB antennas are more desirable because of advantages such as low profile, light weight, low cost, and easy fabrication. Most planar UWB antennas are composed of patches, slots, or stubs of microstrip lines to achieve the wide band of impedance matching [47–51] as shown in Fig. 6.2.

In this chapter, UWB antennas fed by a CPW line are designed, fabricated, and tested. Advantages of a CPW over a microstrip line are low radiation loss, balanced line, low dispersion,

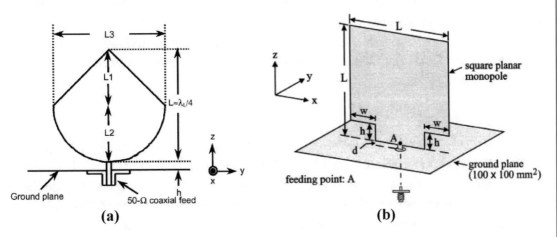

FIGURE 6.1: Various nonplanar UWB antennas whose ground planes are vertical to the radiators. (a) From [45], copyright © IEEE 2004. (b) From [46], reprinted with permission of John Wiley & Sons Inc.

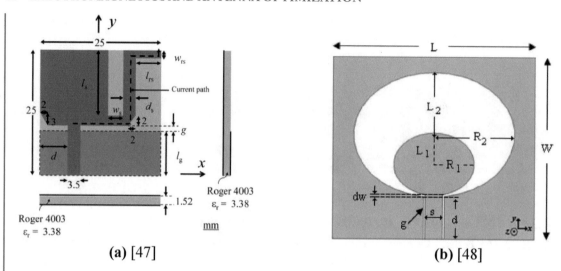

FIGURE 6.2: Planar UWB antennas composed of patches, slots, or stubs of microstrip lines to achieve the wide band of impedance matching. (a) An example composed of patches and microstrip lines; from [47], copyright © IEEE 2007. (b) An example composed of slot geometry; from [48], copyright © IEEE 2006.

coplanar structure, and no vias needed. A full-wave commercial simulator, IE3D [26], along with an external Taguchi's method based optimizer, is used to optimize the UWB antennas. The desired frequency response of the designed antenna is successfully achieved with only a few numbers of iterations. To avoid potential interference between the UWB and WLAN band, a UWB antenna with a band-notched feature at the center frequency (5.5 GHz) of WLAN frequency band is also designed and measured. The results of the optimized antennas show that the impedance bandwidth ranges from 3 to 12.1 GHz, which not only can cover the UWB spectrum but also can be used for X band radar applications.

6.2   A UWB ANTENNA DESIGN

The geometry of a UWB antenna is shown in Fig. 6.3. The antenna is fabricated on the *Rogers RO 4003C* substrate with the dielectric constant $\varepsilon_r = 3.38$, thickness $h = 0.813$ mm, and loss tangent $\tan\delta = 0.002$. The parameter w_f is the width of the CPW line, and g is the gap between the CPW line and the coplanar ground plane. The parameters w_f and g are fixed at 5.0 and 0.25 mm, respectively, to achieve a 50-Ω characteristic impedance. A pair of fork-like microstrip lines is located at the end of the CPW fed line, and the width of two microstrip lines, w_m, is fixed at 2.0 mm. The ground plane size of the antenna is fixed at 55 mm by 55 mm with a rectangular slot in the middle.

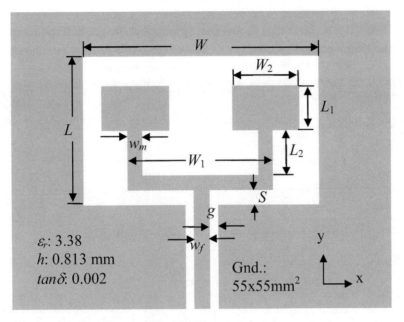

FIGURE 6.3: The proposed geometry of the UWB antenna.

The rectangular slot width and length are W and L, respectively. In IE3D, the antenna is efficiently simulated using the magnetic current modeling method, which assumes an infinite ground plane.

The UWB antenna specification is defined as

$$|S_{11}| < -10 \text{dB} \text{ for } 3 \text{ GHz} < f < 12 \text{ GHz}. \quad (6.1)$$

Seven parameters, W, W_1, W_2, L, L_1, L_2, and S, are optimized to achieve the design goal of the UWB antenna. Note that a pair of metal patches located at the end of the fork-like microstrip lines is used to provide more degrees of freedom for the optimization approach when searching for the optimal solution. The optimization range of W_2 is 2.0–9.0 mm. If the optimized value of W_2 is 2.0 mm, the two stubs would become simply extension parts of the microstrip lines since the width of the microstrip lines is also 2.0 mm. An OA(*18, 7, 3, 2*) [30] is adopted in this UWB antenna optimization. In the optimization process, the initial antenna's dimensions and their optimization ranges are shown in Table 6.1.

The following fitness function is used in the optimization:

$$\text{Fitness} = \int_{3.0}^{12.0} (|S_{11}| + 12)_{dB} \left[\frac{1 + \text{sgn}(|S_{11}| + 12)}{2} \right] df \quad (6.2)$$

TABLE 6.1: The initial dimensions, optimization ranges, and optimized dimensions obtained by Taguchi's method of the UWB antenna case shown in Fig. 6.3 (all dimensions are in millimeters)

PARAMETERS	INITIAL DIMENSIONS	OPTIMIZATION RANGES	OPTIMIZED DIMENSIONS
W	32.2	28.2–40.2	33.38
$W1$	16	10–20	14.07
$W2$	5	2–9	2.0
L	21.1	16.9–24.1	22.57
$L1$	7	4–8.5	4.0
$L2$	5	3–6.5	3.0
S	1	0.4–2	1.26
g	0.25	0	0.25
w_f	5.0	0	5.0
w_m	2	0	2

where the unit of frequency is gigahertz, and df is the frequency interval set to 0.1 GHz. The desired return loss is set to −12 dB, which provides a proper tolerance for the antenna to be designed with the required wideband characteristics. Basically, the fitness can be seen as the difference area where the obtained $|S_{11}|$ exceeds the desired $|S_{11,d}|$. When the return loss is lower than −12 dB, the antenna has a good match and its contribution to fitness is 0. When the return loss is higher than −12 dB, it increases the fitness value. In summary, the smaller value the fitness function has, the better the results obtained towards the desired design goal.

In the optimization process, the converged value is set to 0.01, and the RR is a Gaussian function, i.e., $RR = e^{-(\frac{i}{T})^2}$ where $T = 20$, and the $LD_I(n)$ is increased 1.3 times of $LD_1(n)$. After 37 iterations, the optimization process ends since the termination criterion (3.8) is achieved. The convergence curve of fitness is presented in Fig. 6.4. The dimensions of the optimized antenna are also listed in Table 6.1. The stubs shrink and become microstrip lines due to the fact that the optimized W_2 is 2.0 mm.

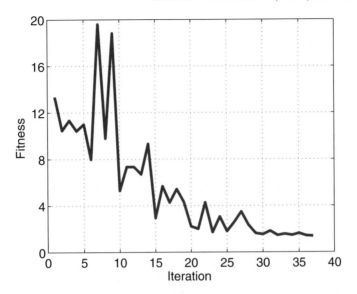

FIGURE 6.4: Convergence curve of the fitness value of the UWB antenna design.

The $|S_{11}|$ of the UWB antenna optimized by Taguchi's method and the $|S_{11}|$ of the initial UWB antenna are shown in Fig. 6.5. It is observed that the desired frequency response of S_{11} is achieved. The simulated $|S_{11}|$ of the initial UWB antenna is quite poor. However, the simulated $|S_{11}|$ obtained from Taguchi's method can cover the frequency band from 3.0 to 13.3 GHz, which corresponds to an impedance bandwidth of 126.4% ($|S_{11}| < -10$ dB). The frequency ratio of upper to lower frequencies is 4.43:1. This situation shows that the proposed approach can successfully optimize the application of a UWB antenna and find the optimal solution.

To validate the optimized performance a UWB antenna is fabricated on the Rogers RO 4003C substrate with the optimized dimensions shown in Table 6.1. A photo of the fabricated antenna is shown in Fig. 6.6. The antenna was measured using an HP 8510C network analyzer. Fig. 6.7 shows the simulated and measured voltage standing wave ratio (VSWR) of the proposed antenna. The measured VSWR can cover the frequency band from 2.8 to 12.1 GHz, which corresponds to an impedance bandwidth of 124.8% (VSWR <2). The frequency ratio of upper to lower frequencies is 4.32:1. Good agreement between simulated and measured frequency responses in the UWB band (3.1–10.6 GHz) is observed. Obviously, the measured bandwidth can cover the UWB band, and the proposed antenna can also be used for applications of X band.

Radiation patterns of the fabricated antenna are measured at 4, 5.5, 8.5, and 10 GHz. Figures 6.8, 6.9, and 6.10 show radiation patterns in the x–z, y–z, and x–y plane, respectively.

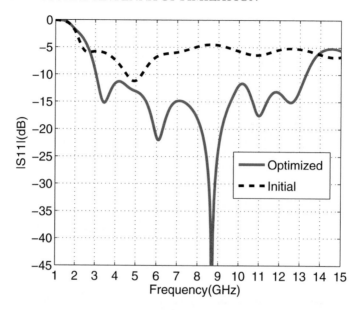

FIGURE 6.5: The $|S_{11}|$ of the UWB antenna optimized by Taguchi's method, and the $|S_{11}|$ of the initial UWB antenna.

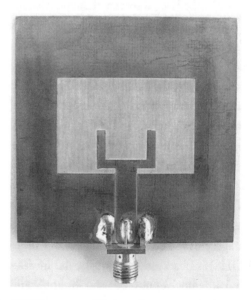

FIGURE 6.6: Picture of the UWB antenna fabricated on *Rogers RO 4003C* substrate. The size of the antenna is 55 mm by 55 mm.

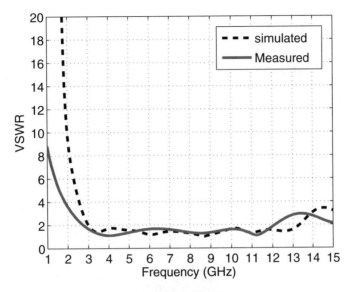

FIGURE 6.7: The simulated and measured VSWR of the UWB antenna.

6.3   A UWB ANTENNA WITH BAND-NOTCH PROPERTY

Since the spectrum of WLAN of *IEEE* 802.11a is located between 5.15 and 5.825 GHz, it is desirable that a UWB antenna has a band-notched feature at the center frequency (5.5 GHz) of WLAN frequency band to avoid potential interference between the UWB and WLAN bands. For slot antennas, conventional methods are used such as adding stubs in a slot [52] as shown in Fig. 6.11. For patch antennas, slots are incorporated on the patches [53–55] to obtain the band-stop feature, as shown in Fig. 6.12. However, the antenna's geometry is changed, and the impedance bandwidth and return loss at other frequencies are affected as well.

In this study, a compact coplanar waveguide BSF shown in Fig. 5.7 is applied to achieve a stop-band between 5 and 6 GHz without changing the antenna's geometry or the frequency response outside the stop-band. This BSF was optimized, and results were shown in the previous chapter. The dimensions of the optimized BSF are used in this study. Note that the substrate and dimensions of CPW-fed lines are identical for both geometries.

The optimized UWB antenna with a compact BSF is fabricated on the *Rogers RO 4003C* substrate as shown in Fig. 6.13. The antenna was measured using HP 8510C network analyzer. Fig. 6.14 shows the simulated and measured return loss of the proposed antenna. Measured results show that a band-stop spectrum from 4.6 to 5.9 GHz is achieved with a center frequency at 5.52 GHz. The measured S_{11} of the antenna with BSF can cover the frequency band from 2.50 to 10.88 GHz

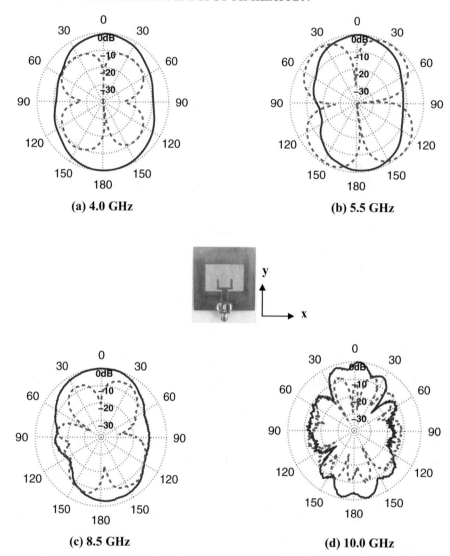

FIGURE 6.8: Measured radiation patterns at 4, 5.5, 8.5, and 10 GHz in the x–z plane (normalized magnitude versus θ). Solid line is copolarization, and dotted line is cross-polarization.

except band-stop spectrum. Reasonable agreement between simulated and measured frequency responses is observed. Obviously, the measured bandwidth can cover the UWB band. The measured return loss curve slightly shifts to a lower frequency while almost maintaining the same frequency bandwidth (around 125%) compared with the measured return loss of the UWB antenna with BSF. However, in the higher frequency part, the −10 dB point of S_{11} decreases from 12.1 to 10.88 GHz

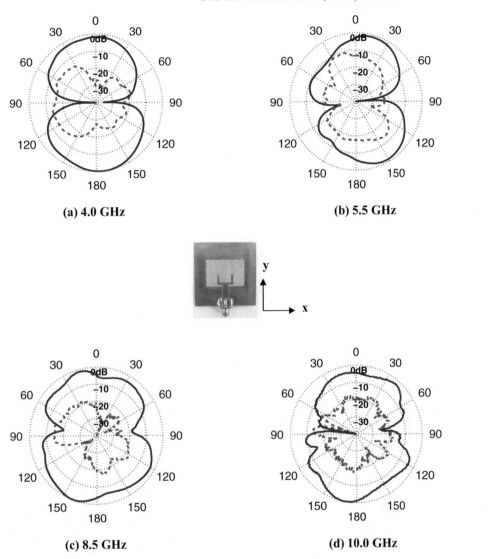

(a) 4.0 GHz

(b) 5.5 GHz

(c) 8.5 GHz

(d) 10.0 GHz

FIGURE 6.9: Measured radiation patterns at 4, 5.5, 8.5, and 10 GHz in the y–z plane (normalized magnitude versus θ). Solid line is copolarization, and dotted line is cross-polarization.

since the second-order resonant mode of the BSF occurs at 11 GHz so that the S_{11} over 11 GHz becomes poor.

Fig. 6.15 compares the computed gains of two UWB antennas over the entire frequency range. The results are calculated in IE3D. The simulated maximum gain is 4 dBi at 3 GHz and the gain increases when frequency increases. A null in the antenna gain is observed at 5.5 GHz for the

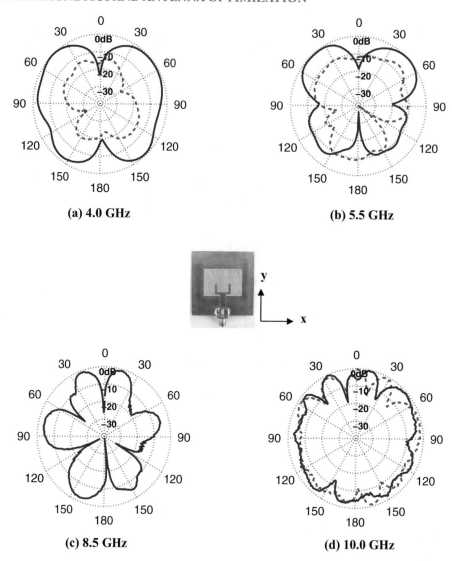

FIGURE 6.10: Measured radiation patterns at 4, 5.5, 8.5, and 10 GHz in the x–y plane (normalized magnitude versus ϕ). Solid line is copolarization, and dotted line is cross-polarization.

UWB antenna with BSF. The maximum gain is reduced significantly to −6 dBi, which is around 10 dB lower than the one without BSF. This result shows that the BSF successfully blocks the spectrum around 5.5 GHz. Moreover, the two curves agree well outside the stop band, which demonstrates that the BSF does not affect other frequency response. Similar observations are noticed in the measured data.

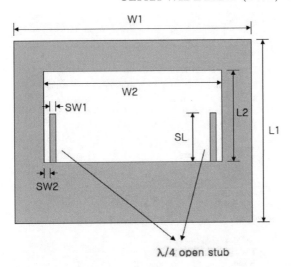

FIGURE 6.11: An example of adding stubs in a slot antenna to obtain band-stop features. From [52], copyright © IEEE 2006.

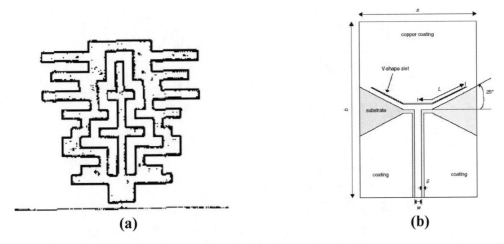

FIGURE 6.12: Examples of cutting slots on the patches to obtain the band-stop feature. (a) From [53], copyright © IET 2007. (b) From [54], copyright © IEEE 2003.

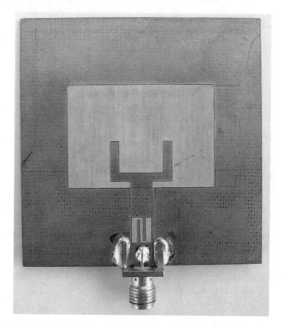

FIGURE 6.13: Picture of the UWB antenna with a compact BSF fabricated on a *Rogers RO 4003C* substrate. The ground plane size of the antenna is 55 mm by 55 mm.

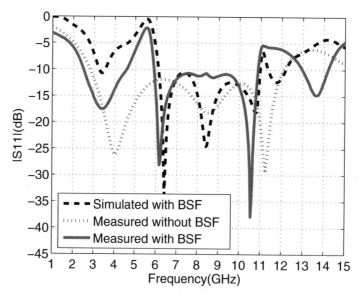

FIGURE 6.14: Simulated and measured return losses of the proposed antenna.

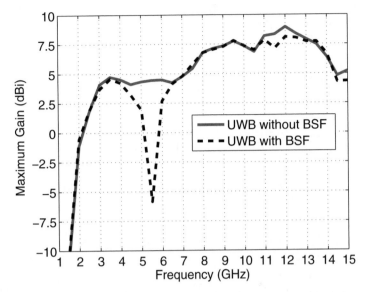

FIGURE 6.15: Comparison of simulated gains of UWB antennas with and without BSF.

•   •   •   •

CHAPTER 7

OA-PSO Method

In previous chapters, Taguchi's method demonstrates its great capability in EM and antenna optimizations. As a key part of Taguchi's method, the OA concept can also be applied in other optimization techniques to improve their performance. For example, this chapter introduces a hybrid method that integrates the OA concept with the PSO technique. We believe the discussion here will create more research opportunities for better optimization techniques.

7.1 CLASSICAL PSO METHOD

The PSO method [18] has been demonstrated to be effective in optimizing difficult, multidimensional, and discontinuous problems in a variety of fields. Like bees searching a field for the location of the highest concentration of flowers, particles in PSO are attracted to the best location found by the entire swarm and to the best location personally encountered by the particle. Eventually, after being attracted to areas of high flower concentration, all bees swarm around the best location. As an evolutionary algorithm, PSO shares the ability of GA to handle arbitrary nonlinear cost functions. One advantage of PSO over GA is the algorithmic simplicity [33].

A flowchart of classical PSO is shown in Fig. 7.1. The entire swarm is composed of N particles, and each particle's position has k dimensions needed be optimized. The optimization process starts from the problem initialization, which defines the solution space, fitness function, initial random positions x_i, and velocities, v_i, of the particles.

When an entire swarm discovers a solution, which is best, the position of the particle is stored in a vector, called G_{best}. When a particle discovers a solution which is better than any it has found previously, the position of the particle is stored in a vector of the particle, called P_{best}. If there are N particles, and each particle has k parameters that should be optimized, the G_{best} is a 1-by-k vector, and P_{best} is an N-by-k matrix.

The velocity of a particle for the next iteration is updated by the following equation:

$$v_{i+1} = w\,v_i + c_1\,\text{rand}\,()\,(P_{best,\,i} - x_i) + c_2\,\text{Rand}\,()\,(G_{best} - x_i), \qquad (7.1)$$

where w is the "inertial weight" and is chosen between 0 and 1.0. Eberhart and Shi [56] suggested that the inertial weight w varies from 0.9 to 0.4 over the course of the iteration. In this research, the

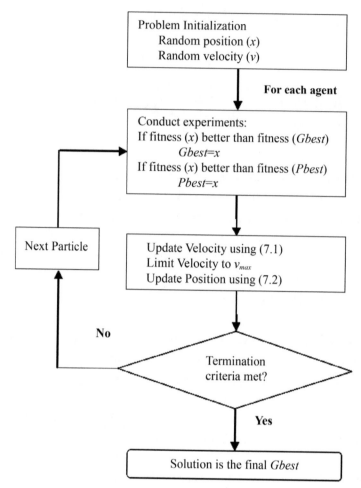

FIGURE 7.1: Flowchart of classical PSO.

inertial weight is decreased linearly from 0.9 to 0.4 over the first 400 runs and $w = 0.4$ is kept after 400 iterations. This setting lets particles have more freedom to search the best position during the first 400 iterations. The parameters c_1 and c_2 are the scaling factors that determine the relative "pull" of P_{best} and G_{best} [33, 57]. Increasing c_1 raises the proportion of P_{best} such that each particle is more encouraged to move toward its own best position. On the other hand, increasing c_2 raises the proportion of G_{best} such that each particle is more encouraged to move toward the current best position. Eberhart and Shi [57] suggested that the best choice for both c_1 and c_2 is 2.0. In addition, v_{i+1} is limited by $\pm v_{max}$, which is the maximum velocity allowed in a given direction. Usually, the value of v_{max} is set to the half size of solution space. The rand() and Rand() are uniform random values that range between 0 and 1.

The new position of a particle for the next iteration is determined according to the following equation:

$$x_{i+1} = x_i + v_{i+1}. \tag{7.2}$$

When x_{i+1} hits the boundary of the optimization range in one of the dimensions, x_{i+1} is set to the bound value of optimization range. Meanwhile, the sign of the velocity is changed, which pulls the particle back toward the solution space.

The optimization process runs iteratively until the fitness of G_{best} meets the termination criterion. It is clear that the PSO is very simple to implement. Its efficiency in EM has been demonstrated in several literatures [33].

7.2   OA-PSO METHOD AND PERFORMANCE COMPARISON

It can be seen that random values offer a wider degree of freedom for particles in the PSO process. However, it is not efficient to initialize the positions of particles using random values since random particles' positions are not uniformly distributed in the entire solution space especially in large dimension problems and the positions may not be located near the optimal solution.

To overcome this drawback, an OA is used to initialize the positions of particles [58]. The basic idea is that an OA can offer fair and balanced distribution in the whole optimization range so that a particle may have a better opportunity to locate near the optimum. As a consequence, the optimization efficiency of the PSO can be improved. This modified PSO is called OA-PSO. Thus, the only difference between the OA-PSO and classical PSO is the initialization process. Instead of random values, the positions of particle are initialized by OA and the procedure is the same as in Section 3.1.2. Note that in this hybrid technique, the number of particles should be equal to the number of rows in an OA.

To compare the efficiency of the classical PSO and OA-PSO, the 10-element, unequally spaced linear array discussed in Chapter 4 is used again as a test example. The same OA (*18, 5, 3, 2*), which is an 18-by-5 array with three-level entries, is used in OA-PSO, and the initialized values of each element are shown in Table 4.2. For a fair comparison, the same number of experiments per iteration is used in classical PSO. Also, all the optimization settings, such as w, c_1, c_2, and converged rate, are kept identical in the two approaches.

Since random values are used in both PSO approaches, the optimization result is not the same in each run. Therefore, 200 independent runs are conducted using each approach and statistical analysis is performed on the obtained 200 samples. The statistical data, such as mean and standard deviation, are used to evaluate the performance of the two optimization approaches.

In this research, two kinds of efficiency comparisons are conducted. In comparison A, the number of iterations is fixed at 350 and the fitness value is recorded at the end of each independent

NUMBER OF PSO PARTICLES: 18	PSO	OA-PSO
TABLE 7.1: The efficiency comparison of two PSO approaches		
Comparison A: fitness, fixed number of iterations at 350		
Mean	0.04547	0.03756
Standard deviation	0.08458	0.07866
Comparison B: number of iterations, fixed fitness value at 0.001		
Mean	583	549
Standard deviation	343	322
200 sample data analyzed for 10-element unequally spaced linear array		

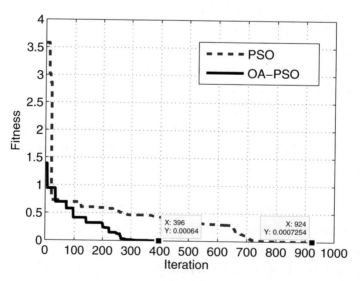

FIGURE 7.2: A convergence comparison of two approaches.

run. The smaller the fitness value, the better the optimization method. In comparison B, the fitness value is fixed at 0.001 as the termination criterion, and then the number of iterations required in each independent run is recorded. A small number of iterations indicates an efficient optimization method.

The statistical results are shown in Table 7.1. The means in OA-PSO are less than those of classical PSO in both comparisons A and B. Therefore, the OA-PSO is better and more efficient than the classical PSO. In addition, the standard deviations of two comparisons in OA-PSO are also less than those of classical PSO, which means that OA-PSO is more robust than classical PSO.

A convergence comparison of two approaches is shown in Fig. 7.2. The termination criterion is set at 0.001 for the same unequally spaced linear array problem. The fitness value of OA-PSO in the first iteration is much smaller than that of classical PSO, which is benefited from the OA initialization. Also, OA-PSO reaches the design goal quicker than PSO. To sum up, the optimization performance of the hybrid OA-PSO method is better than that of classical PSO in this case.

•   •   •   •

CHAPTER 8

Conclusions

8.1 SUMMARY

In this book, a global optimization technique based on Taguchi's method is introduced to the EM community. The concept of OA and the optimization procedure are described in details. The objective of this book was to demonstrate the validity and efficiency of Taguchi's method by optimizing various kinds of antennas and microwave applications. Optimized results show that the desired optimization goals are successfully achieved.

The proposed optimization method is a systematic and efficient algorithm that can solve problems with a high degree of complexity using a relatively small number of experiments in the optimization process. Taguchi's method can deal with a larger number of optimization parameters in a design; thus, it saves much time in a design process. Compared with other approaches such as gradient-based methods and PSO, Taguchi's method is easy to implement and converges to the desired goals quickly. The basic optimization procedure is introduced first, and then followed by several advanced improvement techniques. By increasing the initial level difference, using five-level OA, and adopting a Random-Taguchi's method, one could avoid the optimization process sticking at a local optimum. The optimization performance was also improved by using a Gaussian reduced function which speeded up the convergence of the optimization approach.

Taguchi's method is a good candidate for optimizing both EM and other scientific and engineering problems. In this book, we focus on the applications in EM and antenna engineering. The main contributions are summarized and listed below.

8.1.1 Linear Antenna Arrays

Taguchi's method was used to design three linear antenna arrays. The first antenna pattern possesses nulls in desired directions, which are widely used in smart antenna systems to eliminate the interference from specific noise directions. The excitation magnitudes of a 20-element equally spaced linear array were designed. Optimized results showed that after 23 iterations, the beam width at −40 dB side lobe level is 20.9°, the HPBW is 7.4°, and nulls are below −55 dB in the angle ranges of [50°, 60°] and [120°, 130°], as desired. The fitness value converged to the optimum result quickly.

A sector beam pattern allows the antenna array to have a wider coverage, which exhibits a desired distribution in the entire visible region. To realize the sector beam pattern, we still use a 20-element, equally spaced linear array. However, both excitation magnitudes and phases of the array elements were optimized to shape the antenna pattern. Thus, 20 parameters were optimized simultaneously in the optimization process. This problem is more complicated and can be considered as a high-dimension problem. After 60 iterations, the desired optimum sector beam pattern was obtained. It has a ripple of 0.48 dB, the beam width at −25 dB SLL is 41.2°, and the HPBW is 28.1°. An efficiency comparison between Taguchi's method and PSO was also presented. The reduction in the required number of experiments is around 70%, which shows that Taguchi's method is quicker than PSO to achieve the same optimization goal.

The pattern of suppressed side lobe levels guarantees the radiating or receiving energy to be more focused on the specific directions. A 10-element unequally spaced linear array problem was studied. The element excitations are identical and the element locations were optimized to suppress the SLL. The goal of this optimization was the suppression of the SLL to −18.96 dB, which reduced it by 6 dB compared with that of the equally spaced linear array. The optimization results showed that the SLL was successfully suppressed to −18.96 dB after 60 iterations. The side lobes had approximately the same level. The beam width is 20.7°, SLL is −18.96 dB, and HPBW is 9.8°.

8.1.2   Planar Filter Design

Microwave filters are widely used in telecommunication equipments. In particular, passive planar types of filters are preferable in microwave regions. In this study, Taguchi's optimization method integrated with a full-wave commercial simulator, IE3D, was used to optimize a microstrip BSF, a coplanar waveguide BSF, and a microstrip BPF. The detailed procedure of integrating Taguchi's method with IE3D was presented. The desired frequency responses of the three planar filters were successfully achieved with only a few numbers of iterations, which demonstrated the benefits of this design methodology.

In the coplanar waveguide BSF design, Taguchi's method is also used to determine the main effects of different parameters. The mean of the ranking and the overall ranking were calculated for each parameter to identify the most influential parameters affecting the fitness of the filter.

8.1.3   UWB Antenna Design

Recently, UWB antennas have received great attention in the EM community. Planar UWB antennas are desirable because of advantages such as low profile, light weight, low cost, and easy fabrication. A UWB antenna fed by a CPW line was designed in this study. A full wave commercial simulator, IE3D, along with an external Taguchi-based optimizer was used to optimize a UWB antenna. The desired frequency response of the UWB antenna was successfully achieved with only a few numbers of iterations. Furthermore, to avoid potential interference between the UWB and WLAN band, a UWB antenna with a band-notched feature at the center frequency (5.5 GHz) was also designed.

The optimized UWB antennas were fabricated and measured, and reasonable agreement between simulated and measured frequency responses was observed. The measured bandwidth ranges from 2.8 to 12.1 GHz, which covers both the UWB band (3.1–10.6 GHz) and X band for radar applications (8–12 GHz). For the UWB antenna with a band-notched feature, measured results showed that the S_{11} of the antenna with a BSF can cover the frequency band from 2.50 to 10.88 GHz, which corresponds to an impedance bandwidth of 125.3%. A stop band was achieved with a center frequency at 5.52 GHz. Radiation patterns of the proposed UWB antennas were measured in an anechoic chamber and the data were presented in the book for references.

8.2 FUTURE WORK

As a novel global optimization technique in EM, we believe Taguchi's method will have a profound impact in microwaves and antenna engineering. Some future works are suggested below, which are organized into two areas: algorithm and applications.

8.2.1 Algorithm

This book introduces the basic algorithm of Taguchi's optimization method. Some fundamental questions need to be further addressed. For example, three levels and five levels of orthogonal arrays are used in this book. Each has its own advantages and disadvantages. A further question is whether OA with two or four levels also works or not. In general, what is the guideline of selecting the number of OA levels?

For the examples discussed in this book, the optimization regions for all parameters are continuous space. If a discrete optimization space is presented in a problem, Taguchi's method needs to be modified accordingly. Research development in this area is necessary for many real applications.

Hybrid optimization technique is another potential research direction. In this book, an OA-PSO method is introduced and it shows better performance than the classical PSO method. Integration of Taguchi's method with other techniques will be attempted in the future. The validity and efficiency of hybrid optimization techniques will be carefully evaluated.

8.2.2 Applications

Besides the examples illustrated in previous chapters, Taguchi's optimization method could be further applied in numerous EM problems. Several potential applications are suggested in the list below:

- Thinned antenna arrays. By turning some antenna elements ON or OFF, specific radiation patterns can be realized. Two-level OA may be used in this optimization case.
- Electromagnetic band gap (EBG) structures. Due to the unique EM properties, EBG has been used in many antennas and microwave circuits. A challenging question is to realize compact and broadband EBG designs. Taguchi's optimization method could be a good tool for this design purpose.
- Reflect array antennas. The planar geometry of reflect arrays is desirable in spacecrafts and mobile communication environments. To obtain a broad bandwidth, the element geometry and feed structure should be carefully designed. Many parameters affect the system performance and Taguchi's method could provide an optimum design of these parameters.
- Broadband impedance transformer and matching network. Similar to planar filters, these passive components are widely used in microwave circuits. As a robust and efficient tool, Taguchi's method can be used to optimize their frequency behavior.

In summary, we have an exciting experience during the development and applications of Taguchi's optimization method. We sincerely hope that readers will enjoy this method and find it useful in their own research. Readers comments and suggestions are always welcomed.

· · · ·

Bibliography

[1] R. L. Haupt and S. E. Haupt, *Practical Genetic Algorithms*. New Jersey: John Wiley & Sons Inc., 1998.

[2] J. W. Bandler and S. H. Chen, "Circuit optimization: the state of the art," *IEEE Transactions of Microwave Theory and Techniques*, vol. 46, no. 2, pp. 424–443, February 1998. doi:10.1109/22.3532

[3] V. Rizzoli, A. Costanzo, D. Masotti, A. Lipparini, and F. Mastri, "Computer-aided optimization of nonlinear microwave circuits with the aid of electromagnetic simulation," *IEEE Transactions of Microwave Theory and Techniques*, vol. 52, no. 1, pp. 362–377, January 2004.

[4] Y. Kuwahara, "Multi-objective optimization design of Yagi-Uda antenna," *IEEE Transactions on Antennas and Propagation*, vol. 53, no. 6, pp. 1984–1992, June 2005.

[5] Y. Rahmat-Samii and E. Michielssen, *Electromagnetic Optimization by Genetic Algorithms*. New Jersey: John Wiley & Sons Inc., 1999. doi:10.1109/ICECOM.2003.1290941

[6] G. Taguchi, S. Chowdhury, and Y. Wu, *Taguchi's Quality Engineering Handbook*. New Jersey: John Wiley & Sons Inc., 2005.

[7] H. Nagano, K. Miyano, T. Yamada, and I. Mizushima, "Robust selective-epitaxial-growth process for hybrid SOI wafer," in *IEEE International Symposium on Semiconductor Manufacturing*, San Jose, CA, October 2003, pp. 187–190.

[8] G. Y. Hwang, S. M. Hwang, H. J. Lee, J. H. Kim, K. S. Hong, and W. Y. Lee, "Application of Taguchi method to robust design of acoustic performance in IMT-2000 mobile phones," *IEEE Transactions on Magnetics*, vol. 41, pp. 1900–1903, May 2005. doi:10.1109/TMAG.2005.846255

[9] T. Y. Chou, "Applications of the Taguchi method for optimized package design," in *IEEE 5th Topical Meeting on Electrical Performance of Electronic Packaging*, Napa, CA, October 1996, pp. 14–17.

[10] H. T. Wang, Z. J. Liu, S. X. Chen, and J. P. Yang, "Application of Taguchi method to robust design of BLDC motor performance," *IEEE Transactions on Magnetics*, vol. 35, issue 5, part 2, pp. 3700–3702, September 1999.

[11] A. D. MacDonald, "A modified Taguchi method for the design of broadband spiral cavity absorbers," in *International Symposium on Antennas and Propagation Society*, Dallas, TX, May 1990, vol. 3, pp. 1180–1183.

[12] A. Charles, M. S. Towers, and A. McCowen, "Sensitivity analysis of Jaumann absorbers," *IEEE Proceedings on Microwaves, Antennas and Propagation*, vol. 146, issue 4, pp. 257–262, August 1999. doi:10.1049/ip-map:19990566

[13] J. Felba, K. P. Friedel, and A. Moscicki, "Characterization and performance of electrically conductive adhesives for microwave applications," in *4th International Conference on Adhesive Joining and Coating Technology in Electronics Manufacturing, 2000 Proceedings*, Espoo, Finland, June 2000, pp. 232–239. doi:10.1109/ADHES.2000.860608

[14] P. Gouget, G. Duchamp, and J. Pistre, "Optimisation and comparison of three diplexers based on a new slot to microstrip junction," in *2003 IEEE MTT-S International Microwave Symposium Digest*, Philadelphia, PA, June 2003, vol. 2, pp. 1231–1234. doi:10.1109/MWSYM.2003.1212591

[15] K. L. Virga and R. J. Engelhardt, Jr., "Efficient statistical analysis of microwave circuit performance using design of experiments," in *1993 IEEE MTT-S International Microwave Symposium Digest*, Atlanta, GA, June 1993, vol. 1, pp. 123–126.

[16] W. C. Weng, F. Yang, V. Demir, and Atef Elsherbeni, "Optimization Using Taguchi Method for Electromagnetic Applications," in *The First European Conf on Antennas and Propagation*, Nice, France, October 2006.

[17] W. C. Weng, F. Yang, and A. Z. Elsherbeni, "Linear antenna array synthesis using Taguchi's method: a novel optimization technique in electromagnetics," *IEEE Transactions on Antennas and Propagation*, vol. 55, no. 3, pp. 723–730, March 2007.

[18] J. Kennedy and R. C. Eberhart, "Particle swarm optimization," in *Proceedings of the IEEE Conference on Neural Networks IV*, Piscataway, NJ, 1995. doi:10.1109/ICNN.1995.488968

[19] M. M. Khodier and C. G. Christodoulou, "Linear array geometry synthesis with minimum sidelobe level and null control using particle swarm optimization," *IEEE Transactions on Antennas and Propagation*, vol. 53, no. 8, pp. 2674–2679, August 2005. doi:10.1109/TAP.2005.851762

[20] D. W. Boeringer and D. H. Werner, "Particle swarm optimization versus genetic algorithms for phased array synthesis," *IEEE Transactions on Antennas and Propagation*, vol. 52, no. 3, pp. 771–779, March 2004. doi:10.1109/TAP.2004.825102

[21] D. W. Boeringer, D. H. Werner, and D. W. Machuga, "A simultaneous parameter adaptation scheme for genetic algorithms with application to phased array synthesis," *IEEE Transactions on Antennas and Propagation*, vol. 53, no. 1, pp. 356–371, January 2005. doi:10.1109/TAP.2004.838800

[22] F. J. Ares-Pena, A. Rodriguez-Gonzalez, E. Villanueva-Lopez, and S. R. Rengarajan, "Genetic algorithms in the design and optimization of antenna array patterns," *IEEE Transactions on Antennas and Propagation*, vol. 47, no. 3, pp. 506–510, March 1999. doi:10.1109/8.768786

[23] D. Gies and Y. Rahmat-Samii, "Particle swarm optimization for reconfigurable phased-differentiated array design," *Microwave and Optical Technology Letter*, vol. 38, no. 3, pp. 168–175, August 2003.

[24] Y. B. Tian and J. Qian, "Improve the performance of a linear array by changing the spaces among array elements in terms of genetic algorithm," *IEEE Transactions on Antennas and Propagation*, vol. 53, no. 7, pp. 2226–2230, July 2005. doi:10.1109/TAP.2005.850739

[25] N. Jin and Y. Rahmat-Samii, "A novel design methodology for aperiodic arrays using particle swarm optimization," in *2006 National Radio Science Meeting Digest*, Boulder, CO, January 2006, pp. 69.

[26] *IE3D User's Manual*, Release 11, Zeland Software, Inc., Fremont, CA, February 2005.

[27] FCC, "First Report and Order on Ultra-wideband Technology," *Technical Report*, 2002.

[28] A. S. Hedayat, N. J. A. Sloane, and J. Stufken, *Orthogonal Arrays: Theory and Applications*. Springer-Verlag: New York, 1999.

[29] C. R. Rao, "Factorial experiments derivable from combinatorial arrangements of arrays," *Journal of the Royal Statistical Society*, vol. 9, no. 1, pp. 128–139, 1947. doi:10.2307/2983576

[30] N. J. A. Sloane, "A library of orthogonal array," available: http://www.research.att.com/~njas/oadir/

[31] Y. Leung and Y. Wang, "An orthogonal genetic algorithm with quantization for global numerical optimization," *IEEE Transactions on Evolutionary Computation*, vol. 5, issue 1, pp. 41–53, February 2001.

[32] M. Clerc and J. Kennedy, "The particle swarm—explosion, stability, and convergence in a multidimensional complex space," *IEEE Transactions on Evolutional Computation*, vol. 6, no. 1, pp. 58–73, Feb. 2002. doi:10.1109/4235.985692

[33] J. Robinson and Y. Rahmat-Samii, "Practical swarm optimization in electromagnetics," *IEEE Transactions on Antennas and Propagation*, vol. 52, no. 24, pp. 397–407, February 2004.

[34] M. F. Pantoja, A. R. Bretones, and R. G. Martin, "Benchmark antenna problems for evolutionary optimization algorithms," *IEEE Transactions on Antennas and Propagation*, vol. 55, no. 4, pp. 1111–1120, April 2007.

[35] J. M. Johnson and Y. Rahmat-Samii, "Genetic algorithms in engineering electromagnetics," *IEEE Antennas and Propagation Magazine*, vol. 39, issue 4, pp. 7–21, August 1997. doi:10.1109/74.632992

[36] R. L. Haupt and S. E. Haupt, "Optimum population size and mutation rate for a simple real genetic algorithm that optimizes array factors," *Applied Computational Electromagnetics Society (ACES) Newsletter*, vol. 15, no. 2, July 2000. doi:10.1109/APS.2000.875398

[37] C. A. Balanis, *Antenna Theory: Analysis and Design*. 3rd ed. New Jersey: John Wiley & Sons Inc., 2005.

[38]   R. J. Mailloux, *Phase Array Antenna Handbook*. 2nd ed. Massachusetts: Artech House, 2005.

[39]   J. W. Bandler, R. M. Biernacki, S. H. Chen, D. G. Swanson, and S. Ye Jr., "Microstrip filter design using direct EM field simulation," *IEEE Transactions on Microwave Theory Techniques*, vol. 42, issue 7, pp. 1353–1359, July 1994.

[40]   S. F. Peik and Y. L. Chow, "Genetic algorithms applied to microwave circuit optimization," in *1997 Asia-Pacific Microwave Conference*, December 1997, vol. 2, pp. 857–860. doi:10.1109/APMC.1997.654677

[41]   S. W. Qu, J. L. Li, and Q. Xue, "A band-notched ultra-wideband printed monopole antenna," *IEEE Antennas and Wireless Propagation Letters*, vol. 5, pp 495–498. 2006.

[42]   W. C. Weng, F. Yang, and A. Z. Elsherbeni, "Electromagnetic Optimization Using Taguchi's Method: A Case Study of Band Pass Filter Design," in *IEEE International Symposium on Antennas and Propagation*, Honolulu, HI, June 2007.

[43]   J. S. Hong and M. J. Lancaster, *Microstrip Filters for RF/Microwave Applications*. New Jersey: John Wiley & Sons Inc., 2001.

[44]   Z. N. Chen, "Novel bi-arm rolled monopole for UWB applications," *IEEE Transactions on Antennas and Propagation*, vol. 53, no. 2, pp. 672–677, February 2005. doi:10.1109/TAP.2004.841285

[45]   S. Y. Suh, W. L. Stutzman, and W. A. Davis, "A new ultra-wideband printed monopole antenna: the planar inverted cone antenna (PICA)," *IEEE Transactions on Antennas and Propagation*, vol. 52, no. 5, pp. 1361–1365, May 2004.

[46]   S. W. Su, K. L. Wong, and C. L. Tang, "Ultra-wideband square planar monopole antenna for 802.16a operation in the 2–11 GHz," *Microwave and Optical Technology Letters*, vol. 42, no. 6, pp. 463–466, September 2004. doi:10.1002/mop.20337

[47]   Z. N. Chen, "Small printed ultra-wideband antenna with reduced ground plane effect," *IEEE Transactions on Antennas and Propagation*, vol. 55, no. 2, pp. 383–388, February 2007.

[48]   E. S. Angelopoulus, A. Z. Angelopoulus, D. I. Kaklamani, A. A. Alexandridis, F. L. Lazarakis, and W. A. Davis, "Circular and elliptical CPW-fed slot and microstrip-fed Antennas for ultra-wideband applications," *IEEE Antennas and Wireless Propagation Letter*, vol. 5, pp. 294–297, December 2006.

[49]   J. Kim, T. Yoon, J. Kim, and J. Choi, "Design of an ultra wide-band printed monopole antenna using FDTD and genetic algorithm," *IEEE Microwave and Wireless Components Letters*, vol. 15, no. 6, pp. 395–397, June 2005. doi:10.1109/LMWC.2005.850468

[50]   K. Kkiminami, A. Hirata, and T. Shiozawa, "Double-sided printed bow-tie antenna for UWB communications," *IEEE Antennas and Wireless Propagation Letter*, vol. 3, pp. 152–153, 2004. doi:10.1109/LAWP.2004.832126

[51] R. Chair, A. A. Kishk, and K. F. Lee, "Ultrawide-band coplanar waveguide-fed rectangular slot antenna," *IEEE Antennas and Wireless Propagation Letters*, vol. 3, pp 227–229, 2004. doi:10.1109/LAWP.2004.836580

[52] H. K. Yoon, Y. Lim, W. Lee, Y. J. Yoon, S. M. Ham, and Y. H. Kim, "UWB wide slot antenna with band-notch function," in *IEEE International Symposium on Antennas and Propagation and URSI National Radio Science Meeting*, Albuquerque, NM, July 2006.

[53] Y. Kim, and D. H. Kwon, "CPW-fed planar ultra wideband antenna having a frequency band notch function," *Electronics Letter*, vol. 40, no. 7, pp 403–405. April 2004. doi:10.1049/el:20040302

[54] A. Kerkhoff and H. Ling, "Design of a planar monopole antenna for use with ultra-wideband (UWB) having a band-notched characteristic," in *IEEE International Symposium on Antennas and Propagation Society*, Columbus, OH, June 2003, vol.1, pp. 830–833. doi:10.1109/APS.2003.1217589

[55] A. Kerkhoff and H. Ling, "Frequency notch UWB antennas," in *2003 IEEE Conference on Ultra Wideband Systems and Technologies*, Reston, VA, November 2003, pp. 214–218.

[56] R. C. Eberhart and Y. Shi, "Evolving artificial neural networks," in *Proceedings of the 1998 International Conference on Neural Networks and Brain*, Beijing, PRC, 1998.

[57] R. C. Eberhart and Y. Shi, "Particle swarm optimization: developments, applications and resources," in *Proceedings of the 2001 Congress on Evolutionary Computation*, vol. 1, 2001. doi:10.1109/CEC.2001.934374

[58] Z. Bayraktar, P. L. Werner, and D. H. Werner, "Miniaturization of stochastic linear phased arrays via orthogonal design initialization and a hybrid particle swarm optimizer," in *IEEE International Symposium on Antennas and Propagation and URSI National Radio Science Meeting*, Albuquerque, NM, July 2006. doi:10.1109/APS.2006.1711380

Author Biography

Wei-Chung Weng received the B.S. degree in electronic engineering from National Chang-Hua University of Education, Chang-Hua, Taiwan, in 1993, the M.S. degree in electrical engineering from I-Shou University, Kaohsiung, Taiwan, in 2001, and the Ph.D. degree in electrical engineering from The University of Mississippi, University, USA, in 2007.

Since August 2008, he has been an Assistant Professor in the Department of Electrical Engineering, National Chi Nan University, Puli, Taiwan. From 1993 to 2004, he was a Graduate Research Assistant in the Department of Electrical Engineering, University of Mississippi. From 1993 to 2004 and 2007 to 2008, he was a Teacher in the Department of Computer Science, Kaohsiung Vocational Technical School, Kaohsiung, Taiwan. His research interests include antennas and microwave circuits design, computational electromagnetics, electromagnetic compatibility, and optimization techniques in electromagnetics. He has published more than 20 referred journal and conference papers.

Dr. Weng is a member of the IEEE Antennas and Propagation Society and the Microwave Theory and Techniques Society, and a member of the Institute of Antenna Engineers of Taiwan.

Fan Yang received the Bachelor of Science and Master of Science degrees from Tsinghua University in 1997 and 1999, respectively, and the Doctor of Philosophy degree from University of California, Los Angeles (UCLA) in 2002. From 1994 to 1999, he was a research assistant in the State Key Laboratory of Microwave and Digital Communications, Tsinghua University, China. From 1999 to 2002, he was a graduate student researcher in the Antenna Lab, UCLA. From 2002 to 2004, he was a postdoctoral research engineer in the Electrical Engineering Department, UCLA. He was also an instructor there from 2003 to 2004. In August 2004, he joined the Electrical Engineering Department, The University of Mississippi, as an assistant professor. His research interests include antenna theory, designs, and measurements, electromagnetic band gap structures and their applications, computational electromagnetics and optimization techniques, and radio frequency identification systems. He has published two book chapters and over 90 referred journal articles and conference papers. He is a member of IEEE and was secretary of IEEE AP Society, Los Angeles chapter. He serves as a frequent reviewer for more than 10 scientific journals and book publishers, and has chaired numerous technical sessions in various international symposiums. He was a faculty senator at The University of Mississippi and currently is a member of the University Assessment Committee. For his contributions, he has received several prestigious awards and recognitions. In 2004, he received the Certificate for Exceptional Accomplishment in Research as well as Professional Development Award from UCLA. He was the recipient of Young Scientist Award in the 2005 URSI General Assembly and the 2007 International Symposium on Electromagtic Theory. He was also appointed as The University of Mississippi Faculty Research Fellow in 2005 and 2006.

Atef Z. Elsherbeni received an honor Bachelor of Science degree in electronics and communications, an honor Bachelor of Science degree in applied physics, and a Master in Engineering degree in electrical engineering, all from Cairo University, Cairo, Egypt, in 1976, 1979, and 1982, respectively, and a Doctor of Philosophy degree in Electrical Engineering from Manitoba University, Winnipeg, Manitoba, Canada, in 1987. He was a part-time software and system design engineer from March 1980 to December 1982 at the Automated Data System Center, Cairo, Egypt. From January to August 1987, he was a postdoctoral fellow at Manitoba University. Dr. Elsherbeni joined the faculty at the University of Mississippi in August 1987 as an assistant professor of Electrical Engineering. He advanced to the rank of associate professor on July 1991 and to the rank of Professor on July 1997. On August 2002, he became the director of The School of Engineering CAD Lab and the associate director of The Center for Applied Electromagnetic Systems Research (CAESR) at The University of Mississippi. He was appointed as adjunct professor at The Department of Electrical Engineering and Computer Science of the L.C. Smith College of Engineering and Computer Science at Syracuse University in January 2004. He spent a sabbatical term in 1996 at the Electrical Engineering Department, University of California at Los Angeles (UCLA) and was a visiting professor at Magdeburg University during the summer of 2005. He received the 2006 School of Engineering Senior Faculty Research Award for Outstanding Performance in research, the 2005 School of Engineering Faculty Service Award for Outstanding Performance in Service, The 2004 Valued Service Award from the Applied Computational Electromagnetics Society (ACES) for Outstanding Service as 2003 ACES Symposium Chair, the Mississippi Academy of Science 2003 Outstanding Contribution to Science Award, the 2002 IEEE Region 3 Outstanding Engineering Educator Award, the 2002 School of Engineering Outstanding Engineering Faculty Member of the Year Award, the 2001 ACES Exemplary Service Award for leadership and contributions as Electronic Publishing's managing editor from 1999 to 2001, the 2001 Researcher/Scholar of the year award in the Department of Electrical Engineering, The University of Mississippi, and the 1996 Outstanding Engineering Educator of the IEEE Memphis Section. He has conducted research dealing with scattering and diffraction by dielectric and metal objects, finite difference time domain analysis of passive and active microwave devices including planar transmission lines, field visualization and software development for EM education, interactions of electromagnetic waves with human body, sensors development for monitoring soil moisture, airports noise levels, air quality including haze and humidity, reflector and printed antennas and antenna arrays for radars, UAV, and personal communication systems, antennas for wideband applications, and antenna and material properties measurements. He has coauthored 95 technical journal articles and 21 book chapters, contributed to 250 professional presentations, offered 16 short courses and 18 invited seminars. He is the coauthor of the book entitled Antenna Design and Visualization Using Matlab (Scitech, 2006), the book entitled MATLAB Simulations for Radar Systems Design (CRC Press, 2003), and the main author of the chapters "Handheld Antennas" and "The Finite Difference Time Domain Technique for Microstrip Antennas" in Handbook of Antennas in Wireless Communications (CRC Press, 2001). He is a fellow member of the Institute of Electrical and Electronics Engineers (IEEE). He is the editor-in-chief of ACES Journal and an associate editor to the Radio Science Journal. He serves on the editorial board of the Book Series on Progress in Electromagnetic Research, the Electromagnetic Waves and Applications Journal, and the Computer Applications in Engineering Education Journal. He was the chair of the Engineering and Physics Division of the Mississippi Academy of Science and of the Educational Activity Committee for the IEEE Region 3 Section. His home page can be found at http://www.ee.olemiss.edu/atef, and his email address is elsherbeni@ieee.org.

Printed in the United States
by Baker & Taylor Publisher Services